REFINING NATURE

HISTORY OF THE URBAN ENVIRONMENT

Martin V. Melosi and Joel A. Tarr, EDITORS

REFINING NATURE

STANDARD OIL AND THE LIMITS OF EFFICIENCY

Jonathan Wlasiuk

University of Pittsburgh Press

Published by the University of Pittsburgh Press, Pittsburgh, Pa., 15260
Copyright © 2017, University of Pittsburgh Press
All rights reserved
Manufactured in the United States of America
Printed on acid-free paper
10 9 8 7 6 5 4 3 2 1

ISBN 13: 978-0-8229-6520-6
ISBN 10: 0-8229-6520-8

Cataloging-in-Publication data is available from the Library of Congress

COVER ART: Photomontage of map showing sewage deposits at the mouth of the Cuyahoga River, 1912, courtesy of City of Cleveland, Ohio, and an aerial view of a polluted Cleveland neighborhood, courtesy of the Western Reserve Historical Society, Cleveland, Ohio.
COVER DESIGN: Joel W. Coggins

For Annakiss | Never Surrender

CONTENTS

ACKNOWLEDGMENTS

This book was born out of the rich seminar discussions I participated in at Case Western Reserve University's (CWRU) graduate program in the History of Science, Technology, Environment, and Medicine. The faculty at CWRU demanded the highest level of scholarship but encouraged me to tell this story my own way. Ted Steinberg guided me from the inception of this project just over a decade ago and was never shy to let me know when I was chasing a dead end. Dan Cohen ensured that I spoke to a broader audience, and Peter Shulman saved me from my own excesses. Jessica Green provided a perspective outside the discipline of history and encouraged me to dive deeper into the role of politics in the petroleum industry. My graduate school compatriots—Lyz Bly, Sam Duncan, Jon Hazlett, and Ben Sperry—spent countless hours in seminar rooms, cafes, and bars helping me shape the narrative that follows. At CWRU, this project benefited from the financial support of the Baker-Nord Center for the Humanities and the Encyclopedia of Cleveland History.

I also received invaluable criticism from the greater community of environmental historians. John McNeill championed my work and helped me negotiate the perils of converting a dissertation into a proper book manuscript. Brian Black and Christine Rosen offered valuable criticism on conference papers stemming from my manuscript. Lisa Brady, as editor of the journal *Environmental History*, made me appreciate the value of a diligent editor. My footnotes were a dumpster fire before Alex Wolfe and the copy editors at the University of Pittsburgh Press cleaned them up. Christopher Jones, Nic Mink, Alan Roe, and Edmund Russell steered me out of editing doldrums with new perspectives. Sandy Crooms has a generous heart and a knack for making sense out of chaos. It has been an honor to receive feedback from Martin Melosi and Joel Tarr for a book modeled on their own work.

I was fortunate to work with passionate archivists who were willing to invest time and effort in my personal quest. Ann Sindelar is the lorekeeper of the Western Reserve Historical Society and often brought sources to my attention that I never knew existed. Steve McShane has dominion over the Calumet Regional Archives and is desperate to share it with scholars. Tom Rosenbaum and the late Ken Rose opened the gates of the Rockefeller

Archives Center to me and let me know when donuts were in the kitchen. I also owe my gratitude to the staff of the Briscoe Center for American History on the campus of the University of Texas at Austin and the anonymous referees who provided critical comments on my article and book manuscripts.

My family, at times, felt as though they shared joint custody over me with this project. My wife, Annakiss Mauser-Martinez, patiently aided me in my most important discovery: the balance between work and family. My father, Walter Wlasiuk, came to this country as a refugee from a devastated Europe at the end of the Second World War. His grandsons, Aleksander and Sebastian, are a living reminder to me that although the past structures our present, the future remains unwritten.

REFINING NATURE

INTRODUCTION

Although no corporation currently operates under the name Standard Oil Company, the old petroleum trust haunts the present. Despite a recent dip in oil prices, two of the top three corporations on the Fortune 500 list in 2015 are descendants of the Standard Oil Trust. Millions of Americans fill their gas tanks at stations operated by corporations that began under the Standard Oil name. Conoco, Marathon, Exxon/Mobil, and Chevron all constituted branches of Standard Oil before the Supreme Court, in 1911, sundered the trust into thirty-four separate companies. Once the largest refinery in the world, the massive industrial complex in Whiting, Indiana, operated by British Petroleum continues to refine crude oil 128 years after Standard Oil began construction. In the closing years of the nineteenth century, economists, historians, and biographers began to publish what has become a sizable library of books seeking to explain the company's command of a resource that would fuel the modern world. Standard Oil, as many previous studies argue, crushed competitors through superior organization, efficient refining techniques, and outright corruption. The Standard

Oil Company conquered the petroleum trade by manipulating both nature and politics but the success of the company met material limits that explain the tarnished image of the petroleum industry today. In a 2016 poll conducted by Gallup only 37 percent of Americans had a positive view of oil and gas companies, ranking them even below the banking industry. How did we come to distrust the very people that make our modern lives possible? I believe the answer lies in the business model formed during the earliest years of the oil industry. As Standard Oil refined crude oil into a spectrum of commodities, it also "refined" the classical elements (Earth, Fire, Water, Air) into factors of production without regard to their environmental context. The results would produce a series of environmental injustices upon communities that reverberate to this day.[1]

Most students of history learn about the Standard Oil Company when their American history courses reach the Gilded Age, that loosely defined time period shoehorned in somewhere between the Civil War and the Progressive Era. Standard Oil and its shrewd architect, John D. Rockefeller, serve as a convenient illustration of the economic power of integrating every level of production into a single business. Students learn how the control of a resource from its source in nature to its final form as a commodity reduces costs, empowering a corporation to undercut competition and gain market share. If the company saves enough in production and maintains high quality, it can capture a market and emerge as a bona fide monopoly. Some corporations took this model to extremes, founding company towns that offered workers company grocery stores, company housing, and (often) a company government. Henry Ford famously established a company town in the heart of the Amazon to produce his own rubber, and George Pullman built Pullman, Illinois, just south of Chicago. Standard Oil founded its own company town—Whiting, Indiana—in the last decade of the nineteenth century around the largest refinery in the world. The absolute control over workers and consumers made possible by company towns and monopoly seem more akin to feudalism than the supposed free markets of capitalism. Although many histories hail Rockefeller as the consummate capitalist, he believed "the public are not benefitted by competition." Instead, Rockefeller envisioned an efficient utopia controlled by corporate conglomerates. In his own autobiography, published in 1909, he foreclosed on any alternative: "It is too late to argue about advantages of industrial combinations. They are a necessity."[2]

John D. Rockefeller sought to organize the environment with a missionary zeal from his earliest days. As a child growing up amid the farms of upstate New York, he once cut down a tree before breakfast because it "interfered with the view from the windows of the dining-room." Standard Oil became an instrument for his desire to improve, or refine, nature. The company integrated control of America's petroleum supply chain by first focusing on adding value to crude oil through superior refining techniques. Within twenty years, however, the company owned producing wells, tank car and pipeline companies, and distributed kerosene to a global network of sales agents. At the height of its power in the 1880s, Standard Oil controlled 90 percent of the American kerosene trade. Rockefeller believed that the principles of scientific efficiency, or the cutting of costs through vertical integration and strict management, should extend beyond business to individual behavior. "I shall hail the day," Rockefeller admitted in an interview, "when our watchword shall be efficiency, as applied to labor, to people in all positions."[3]

Although it created competitive advantages in the Gilded Age, the strategy of vertical integration confronted environmental and social barriers in the twentieth century that favored the "outsourcing" of resource production and transportation. This history of Standard Oil reveals why outsourcing various levels of production (what Bartow J. Elmore has called "Coca-Cola Capitalism") became cheaper and politically convenient. Standard Oil, local governments, and citizens struggled to control the environmental consequences of drilling, transporting, refining, and selling petroleum. The corporate utopia of the company town collapsed under the weight of its own internal contradictions. The mass consumption and production at the heart of the twentieth-century consumer economy demanded an equally powerful "mass destruction" of environments, according to environmental historian Timothy LeCain. Here lies the dark side of scientific efficiency, for the same practices that squeezed every cent of profit from nature also streamlined the production of wastes. Refineries filled rivers and lakes with caustic acids and waste oils that killed fish, disintegrated the hulls of ships, and threatened public drinking water. Burning coal and oil to fire refinery boilers created a pall of smoke and soot so toxic that the city government of Cleveland, nicknamed the "Forest City," somberly recorded the death of its last virgin trees at the dawn of the twentieth century. Kerosene lamps spontaneously combusted, creating a new incendiary landscape. Industrial production, in short, produced cascading effects within nature that blurred elemental lines as rivers caught fire, soot became airborne, and sickness infested some of the richest cities in human history. Standard Oil did not account for these environmental effects because they transcended property boundaries, which served as the horizon of their worldview. This book shows what happened when nature did not fit into these standardized plans; when efficiency confronted ecology. The regulatory world that emerged in the second half of the twentieth century was, in part, a response to the social and environmental disaster of Rockefeller's dream confronting environmental limits.[4]

The word "limits" in the subtitle of this book has invited criticism. The field of environmental history emerged in the decade after the Club of Rome commissioned the report *The Limits to Growth* in 1972 when scholars gravitated to stories of environmental decline. Some historians today now find such "declensionist" histories passé, and a few argue that stories of decline exist only in the minds of activist scholars. The science writer Leigh Phillips has characterized declensionist history as "Collapse Porn." My only response to such critics lies in the archives that hold the primary documents and in the ecosystems surrounding Standard Oil's refineries that constitute the evidence base of this study. A mountain of research has documented the business success of John D. Rockefeller and the innovative organization of Standard Oil Company. I do not refute this scholarship. I visited the same archives that served as the foundation for these classic histories of the company and its enigmatic creator, but the environment was a blind spot for most of these works. The purpose of this history is to place the rise of the Standard Oil Company in the context of the declining environmental stability that led to the federal environmental legislation of the 1960s and 1970s.[5]

The time frame for this study, the century following the Civil War, witnessed a remarkable experiment in corporate autonomy. Although the US Supreme Court abolished the Standard Oil Trust in 1911, this decision only prohibited the thirty-four separate corporations within the trust from acting in concert. The 1911 decision did not fundamentally alter the industry's relationship with nature or society. Instead, the Standard Oil Company's offspring set the terms of those relationships with little change from the Gilded Age until a wave of state and federal environmental legislation passed in the 1960s and 1970s. As such, Standard Oil enjoyed a "long" Gilded Age that stretched deep into the twentieth century. As Standard Oil's sundered parts rejoin (most significantly in the 1999 Exxon-Mobil merger), the public checks on corporate power appear tenuous and short-lived when compared to the corporation's environmental and social legacy.

The human and corporate descendants of John D. Rockefeller have done quite well for themselves since his death in 1937. The communities bordering Standard Oil refineries have followed a different trajectory. Although Rockefeller's oil empire would encompass the globe by the end of the nineteenth century, I focus on the cradle of the corporation along the Cuyahoga River in Cleveland, Ohio, as well as the company town created at Whiting, Indiana. With perennial appearances on Forbes's list of "America's

Most Miserable Cities," Cleveland has become the poster child of deindustrialization. In a viral video filmed in the aftermath of the 2008 financial crisis, local comedian Mike Polk Jr. created a satirical tourism advertisement documenting classic rustbelt scenes of urban poverty and crumbling infrastructure. In Polk's video, the cheery jingle "Here's the place where there used to be industry!" accompanies a close-up of Cleveland's flats district where Standard Oil gave birth to the petroleum economy. Despite efforts to resurrect the city's image, Cleveland remains the "Mistake by the Lake." It is an easy punch line communicating civic and industrial failure, a place so dysfunctional and polluted that its river caught fire. Students of history are stunned to discover that this same landscape gave birth to the modern petroleum industry and served as the launchpad for the career of John D. Rockefeller, who would amass one of the greatest fortunes in human history. The "Mistake by the Lake" earned a more flattering reputation during the nineteenth century as the "Forest City," built on the wealth of abundant natural advantages. Rockefeller and his closest associates at Standard Oil lived downtown on "Millionaires' Row," a several-miles-long strip of palaces built on the industrial wealth of the city. What went wrong in the following hundred years?[6]

This book is organized according to Standard Oil's relationship to four elements: earth, fire, water, and air. Chapter One examines how changes in the way Rockefeller viewed and used the land played a significant role in the success of his oil empire. Standard Oil became a central actor in the critical shift from animal to machine power at the heart of the industrial revolution, completely reorganizing the relationship between Cleveland and the surrounding environment. Rivers were made straight and deep, roads were replaced with railroads and later pipelines, and Cleveland became the heart of a technological system that converted petroleum into kerosene by employing the most efficient, cost-conscious methods available.[7]

Chapter Two examines how the process of converting rock oil to kerosene created a new flammable landscape that proved hazardous to the company's workers and to consumers of Standard Oil's products. Even without human intervention crude petroleum emitted flammable vapors, and the early oil fields of western Pennsylvania became infamous as one of the most combustible landscapes during the first oil boom. By refining crude into volatile fractions including gasoline, Standard Oil's refineries became a minefield of explosive liquids where a lit match, static spark, or spontaneous combustion could convert passive nature into an explosive shower of fire. Standard Oil pushed the boundaries of safety when price wars with competitors reduced supplies of high-quality illuminants, which caused some sales agents to mix more volatile oils into their kerosene stock. The combustive properties of petroleum embodied the devil's bargain of industrialism: the same fire that illuminated a dark dwelling came with explosive new risks.[8]

Chapter Three reveals how Standard Oil altered Cleveland's relationship to water. In the 1850s the city of Cleveland chose to modernize its water supply, moving from a decentralized system based upon local wells and springs to a centralized waterworks system that provided the population with clean water from Lake Erie. Technology, it appeared at the time, had rescued the city from its environmental circumstances. In concentrating their source of drinking water to a single opening, the city of Cleveland wagered that the eight hundred feet of water between the waterworks crib and the mouth of the Cuyahoga River would serve as a better environmental buffer than the decentralized system of wells it replaced. The gamble paid off for a full decade, until the Cuyahoga River connected the industrial effluent of John D. Rockefeller's nascent petroleum business to the drinking taps of the entire city. This represents perhaps the most startling illustration of Rockefeller's zeal to eliminate waste and increase efficiency. The company dumped thousands of barrels of gasoline, sulfuric acid, and petroleum sludge into the Cuyahoga so as to eliminate the burden from Standard Oil's ledger books. While this eliminated costs for the company—and avoided the construction of storage tanks, the payment of higher insurance rates, and the invention of a market for their use—it reveals how the efficiency and cost savings of one corporation must be understood in the context of its effect on the health of the entire community, including the surrounding environment.

Chapter Four examines the problems created for Cleveland as refineries filled the city's atmosphere with the by-products of fossil fuel combustion. The industrial poisons released into Cleveland's air profoundly altered the city's growth and created a political conflict between advocates seeking government regulation of industry and defenders of laissez-faire during the last decade of the nineteenth century. Many histories of industrialism have noted the early positive association of air pollution with prosperity. Clevelanders seem to have developed a different aesthetic. Refining petroleum released carbon monoxide and sulfur dioxide into the atmosphere, choking workers and destroying plant life. As early as 1870, the air that blew off Lake Erie was filled with the "rancid, suffocating breath of the oil tanks," according to the editor of the *Cleveland Leader*, who found the smoke "so dense in the lower part of the city that persons whose lungs are at all delicate find it next to impossible to sleep." Cleveland's affluent residents fled to suburban settlements, often of their own creation, where they could enjoy clean air. By the opening years of the twentieth century, air pollution had obliterated most of the trees in the "Forest City" including its last stand of old-growth forest. Although Standard Oil endured periodic inspections and petty fines, Rockefeller's desire to "pay a profit to nobody" included the atmosphere, which absorbed the untallied costs of the kerosene industry.[9]

Chapter Five examines the company town that Standard Oil established in Whiting, Indiana, on the southern shore of Lake Michigan. JDR, frustrated by the environmental and social limits on his power in Cleveland, entered the quiet farming and fishing community of northwest Indiana and molded the environment and society to align with the goals of his corporation. Whiting served as a model for the industrial suburb of the early twentieth century. The company drained and canalized the marshland of northwest Indiana, replacing a diverse ecosystem with the largest oil refinery on the planet. The new community, made up almost exclusively of Rockefeller's employees, elected Standard executives to positions of political power, effectively eliminating the threat of any democratic check on his empire. When land became scarce, Rockefeller turned to engineering technology to create new shoreline, at places extending several miles into Lake Michigan and altering what had been a stable ecosystem. This chapter reveals how Standard Oil embodied the self-destructive tendencies of modernist schemes by pushing the environmental costs of kerosene production onto the commons until they reached as far away as Chicago and sparked federal action, effectively ending the company's "long" Gilded Age.[10]

Following the documentary path of Standard Oil in the kerosene age has not been easy. Every scholar must confront the selection bias of the manuscript collections donated by the company and its agents, but prying into the operation of one of the most secretive corporations of the Gilded Age presented unanticipated challenges. Even the Cleveland Board of Trade, which claimed Rockefeller as a member, struggled to compile even basic statistics on Standard Oil. "Unfortunately," the board of trade reported, "the officers of the company will give no detail regarding these works." More problematic for modern historians has been the disappearance of entire collections. In her research on Standard Oil's reach into Europe, Alison Frank followed the footnotes of Ralph Hidy and Muriel Hidy's *Pioneering in Big Business* (1955), a classic business history of Standard Oil, to the Briscoe Center for American History at the University of Texas. "According to the archivist responsible for the Exxon/Mobil papers at the University of Texas," Frank learned, "the papers referenced by the Hidys have since disappeared." My own search for internal documents encountered similar hindrance. My primary research for Chapter Five on the company town of Whiting, Indiana, started initially by searching for the records of Standard Oil of Indiana. This was complicated by the corporation's restructuring into American Oil Company (AMOCO) and its purchase more recently by British Petroleum. When I contacted BP's archive department at the University of Warwick, they informed me that no such records exist (although local journalists and historians insist that the company maintains meticulous care of its documentary history in a climate-controlled iron mine in the Taconite region of Minnesota). Despite these challenges and lacunae, manuscript collections across the United States contain ample evidence in city reports, sanitation journals, lawsuits, newspapers, and some internal documents that demonstrate how urban environments experienced the kerosene age.[11]

The Gilded Age represents an ecological watershed in the history of the United States. Americans had opened the nineteenth century in awe of the seemingly endless continent before them. By the close of the century, however, they grew anxious of despoiling the dwindling natural wealth they had acquired. Along the way they felt, in the words of Henry Demarest Lloyd, they had favored "Wealth Against Commonwealth" and turned their backs on values that had distinguished them from their corrupt European homeland. The Gilded Age as an American cautionary tale of unregulated capitalism is the subject of a troubling revision in recent years. Spearheaded by libertarian think tanks, scholars are attacking the historical foundation of environmental regulations that emerged because of the practices of the Standard Oil Company and other industrial corporations. Jonathan Adler, a legal scholar and proponent of "free market" environmentalism, has argued our "federal command-and-control" regulations, such as the Clean Water Act, rest on a historical "fable" of environmental decline. As communities across the United States find themselves confronted with the promise of economic salvation in form of a "fracking" boom, the past offers us a reminder that the wealth promised in a boom never sets roots as deep as the environmental and social consequences it creates. Millionaires' Row eventually gives way to America's Most Miserable City. We should respect the legacy of entire generations of Americans who lived in cities where the air made you vomit, tap water contained gasoline and petroleum acids, and kerosene lamps became bedside firebombs. This was no fable. It was the expression of the limits of the efficiency game at the heart of the Standard Oil Company.[12]

ONE

IMPROVED EARTH

The 1830s represent an important crossroads in the history of the American economy. Freed from the restrictions imposed by the British imperial relationship that relegated the colonies to a producer of raw materials and consumer of finished British goods, American entrepreneurs turned their energy to refining the North American environment. In a conscious effort to import the Industrial Revolution, American capitalists reorganized labor and natural resources to secure their independence from European manufactured goods. Born in 1839, John Davison Rockefeller entered this changing landscape in the same decade that saw the birth of fellow industrial giants Andrew Carnegie (1835), Jay Gould (1836), J. P. Morgan (1837), and Philip Danforth Armour (1832). These men did not emerge from a vacuum. The cornerstone for their future empires had already been laid by Thomas Jefferson's Land Ordinance of 1785, which transformed the young nation's territory into a checkerboard of legally imperishable property. Americans quickly integrated this land into a growing commodity marketplace. As Western historian Patricia Nelson Limerick has argued, the push toward developing the nation's natural resources constituted "not a simple process of territorial expansion, but an array of efforts to wrap the concept of property around unwieldy objects." These "unwieldy objects," whether in the form of veins of iron and coal, animals, or subterranean oil reserves, in fact, all of nature, became mere commodities for the cohort of young industrialists born into a culture defined by acquisitive capitalism. An environmental history of Standard Oil is ultimately the story of establishing control over nature through the creation of a new legal and economic framework. Through incorporation of the Standard Oil Company, John D. Rockefeller constructed a legal structure for the private capture of nature that had begun with the Land Ordinance.[1]

Various cultural, environmental, and economic conditions structured the personality of JDR and the industrial organization he created. The industrial transformation of petroleum created unintended consequences, which challenged the republican faith that economic development created economic and social independence. What does it mean when economic freedom produces a monopoly so powerful it swallows all of its competitors? How do you weigh the freedom of refineries to fill the air with choking soot against the freedom of downwind communities to breathe clean air? Rockefeller made his peace with these conflicts by favoring efficiency over freedom, steering the American economy, if not the culture, away from its democratic roots. "The day of individual competition in large affairs is past and gone," Rockefeller wrote in his autobiography, "you might just as well argue that we should go back to hand labour and throw away our efficient machines." The application of scientific methods to business raised productivity and managerial control, replacing the disorganized but rich economic environment that gave birth to a generation of entrepreneurs. In this sense, Rockefeller foreclosed on the possibility of others following his footsteps, at least in petroleum.[2]

In his history of Gilded Age America, Alan Trachtenberg recognized this "major realignment of economic power" from acquisitive republicanism to corporate capitalism as a watershed in American economic history. In a single generation, a rash of corporate mergers concentrated American economic power in fewer and fewer hands, so that by the first decade of the twentieth century the largest three hundred corporations controlled about 40 percent of *all* manufacturing in the country. The trend would continue into the future so that by 2010 the top three hundred corporations would control 90 percent of all profits in the American economy. Before JDR could achieve such mastery in his own industry we must understand the origin of his ideas about labor, wealth, and nature as well as the history of the landscapes he would refine to his needs. This is a story about a man and corporation but it begins and ends with the land.[3]

EXODUS

By the time Moses Cleaveland scrambled up the banks of the Cuyahoga to establish the city of his namesake in 1797, the land bordering Lake Erie had become a vast demilitarized zone, long stripped of its valuable beaver pelts and resting between three colonial powers and displaced Amerindian communities. Between the Iroquoian "Mourning Wars" of the seventeenth century and significant white settlement in the first quarter of the nineteenth century, the land bordering Lake Erie enjoyed a resurgence in biological diversity. The fur trader and British agent George Croghan reported in the mid-eighteenth century that the country on the banks of the Ohio and Scioto rivers "abound[ed] with buffalo, bears . . . and all sorts of wild game in such plenty, that we killed out of our boats as much as we wanted." Charged with making peace with the Susquehanna refugees that poured into Ohio country following a century of war and disease, Croghan's party took comfort from a storm at the mouth of the Cuyahoga River and "found several Indians of the Ottawa Nation hunting" the land's ample game.[4]

Such natural abundance made Ohio country a prize for the new American republic. The hunger for Native land beyond the British Proclamation Line on the Appalachian ridge was one of many factors that propelled colonists to sever their bond with England. Benjamin Franklin believed that the rich environment of the Great Lakes, if settled by enterprising pioneers, would form the foundation of a new American empire. His utilitarian descriptions of the land prefigured the material capitalist perspective of Rockefeller.

> The great country back of the Apalachian [*sic*] mountains, on both sides of the Ohio, and between that river and the lakes; is now well known both to the English and French, to be one of the finest in North America, for the extreme richness and fertility of the land; the healthy temperature of the air, and mildness of the climate; the plenty of hunting, fishing, and fowling; the facility of trade with the Indians; and the vast convenience of inland navigation or water-carriage by the lakes and great rivers many hundred of leagues around. From these natural advantages it must undoubtedly (perhaps in less than another century) become a populous and powerful dominion; and a great accession of power, either to England or France.

To secure its new inland empire for the American republic, Franklin believed two forts should be erected—one "at the mouth of the Hioaga [Cuyahoga], on the south side of lake Erie, where a port should be formed, and a town erected, for the trade of the lakes."[5] Franklin studied nature in the hopes of eliminating human dependency on each other, which early republicans believed would poison a democracy of citizens who would vote with their stomachs instead of their cultivated ideals. The early republican zeal to exercise power over the continent embedded the utilitarian conception of nature into the creation myth of the American Midwest.[6]

The settlers who followed Moses Cleaveland into the promised land and purchased land from the Connecticut Land Company searched in vain "for the extreme richness and fertility of the land; the healthy temperature of the air, and mildness of the climate" that Benjamin Franklin promised. The historian Robert A. Wheeler has gathered many of the earliest written descriptions of the land in Ohio Territory in his book *Visions of the Western Reserve*. For example, an early settler to the Western Reserve reported "for four Weeks our people who kept about did little else than take care of the sick." Henry Leavitt Ellsworth, the first US Commissioner of Patents, took away an unfavorable view of the Cuyahoga River. The Cuyahoga proved "not navigable but with small boats." He added, the general "unhealthyness of Cleaveland will be a great hindrance to its settlement." The mouth of the Cuyahoga, believed by many to be the perfect location for settlement for its access to lake and river trade, was a tangled mess of driftwood, sandbars, and decomposing plants. John Melish, a Scottish traveler who visited the struggling city of Cleaveland in 1811, reported the "mouth of the river is choaked [*sic*] up by a sand-bar, which dams up the water." The Cuyahoga, he continued, "stands in a deep pool, two or three miles long; and the water being stagnant, and contaminated by decaying vegetables, afflicts the inhabitants on its margin with fever and ague. . . . for the smell was almost insufferable." These circumstances reinforced utilitarian notions of perfecting an inefficient nature by reorganizing the environment and engendered a belief among the early settlers that nature's order served as an obstacle to their plans.[7]

Cleaveland grew from its meager early population to 606 souls in 1820. By the mid-1820s, a recent emigrant observed the rough-hewn appearance of the frontier city, noting "the Public Square was begemmed with stumps," recently cut to create "its crowning jewel, a log courthouse," while "[t]he eastern border of the Square," the hopeful pioneer remembered, "was skirted by the native forest which abounded in rabbits and squirrels and afforded the villagers a 'happy hunting ground.'" In the 1830s, when the city's name took on its current form, Cleveland's population nearly doubled to 1,075 and supported two newspapers, several churches, and a public market. This influx of population to Ohio elected a growing number of representatives to the US Congress, who in turn made the dreams of Franklin and Washington a reality by approving legislation to dredge the Cuyahoga of its snags and sandbars, create a safe harbor for lake transportation, and construct the Ohio and Erie Canal to connect lake and river commerce to oceanic markets. Ohioans began to refine nature well before John Rockefeller drew his first breath.[8]

The canal connecting the Cuyahoga with the Ohio River brought labor into the service of the developing Atlantic market. Work began on the canal in 1825, and within two years the canal linked the navigable mouth of the Cuyahoga at Cleveland with the booming mill town of Akron, forty miles to the southeast. By 1832 merchants could ship their wares from Cleveland, through Akron and on to the Ohio River, giving them access to trade with the growing nation's interior from Cincinnati to the mouth of the Mississippi. William Bullock traveled from New Orleans to New York entirely by river in 1827 and found the canal path "already fully developed, in its whole line, crowded with boats of considerable size, laden with the various produce of the western and northern states." This link with the Atlantic economy served as a catalyst for further growth of Cleveland, perfectly positioned at the intersection of Lake Erie and the canal path. Bullock noted in his journal a swarm of emigrants seeking the new opportunities created by the canal. "It was really surprising to see the number of poor emigrants," wrote Bullock, "thus proceeding to their destination (many of them were Irish, and on their way to the Ohio), induced to try their fortune with their countrymen."[9]

ORGANIZING NATURE

How do we develop our value systems? Are they merely a product of our culture, absorbed through contact with our community of friends and family? John D. Rockefeller's early psychological development has been vividly reconstructed by Ron Chernow in his 1998 best-selling biography *Titan*, but it is worth taking a moment to place this maturation period in an environmental context. John entered the bustling frontier market after it had already been transformed by a generation of infrastructure improvements and an exploding population. His family contributed to the growing population by migrating to Cleveland from the religiously charged "burned-over" district of western New York in the early 1850s. John's mother, Eliza Davison, turned away from the material world and imbued her son with the stern spirituality of a Baptist raised in the aftermath of the Great Awakening. John's father, William Rockefeller, provided the materialist yin to Eliza's spiritual yang. Before his move to Cleveland, young John learned about finance by managing his family's meager wealth. William Rockefeller eschewed the new wage labor economy by making a living as a peddler in a frontier version of the modern "traveling salesman." "Big Bill" won marksman contests, hawked questionable botanical medicines, faked his way into people's pockets by posing as a deaf-mute, and (some biographers contend) married John's mother in a bid for her family's wealth. His commercial jaunts separated him from the family for extended spells, but he groomed John in money matters. Bill sent John to purchase wood for the family house and imparted an early sense of economy. "My father told me to select only the solid wood and the straight wood and not to put any limbs in it or any punky wood," John recalled later in his life. Bill gave his children lessons in bookkeeping and imparted these teachings in his own idiosyncratic way, allegedly claiming that "I cheat my boys every chance I get. I want to make 'em sharp."[10]

John absorbed these lessons and employed them at an early age. "To my father I owe a great debt in that he himself trained me to practical ways," Rockefeller remembered in his autobiography. He expanded on the lessons imparted from father to son, "he taught me the principles and methods of business. From early boyhood I kept a little book which I remember I called Ledger A—and this little volume is still preserved—containing my receipts and expenditures as well as an account of the small sums that I was taught to give away regularly." Young John bought candy in bulk and sold it individually at a modest profit, began saving his overhead, and at age seven stole eggs from a wild turkey nest and began a small but profitable livestock trade. As a child growing up in the Finger Lakes region of New York state, John enjoyed the outdoors, playing baseball and even sneaking out of the house late at night to skate on the Susquehanna River or Lake Owasco. As a child of the country, John also displayed an instrumentalist understanding of the natural world. "I remember when I was hardly more than a boy," Rockefeller wrote in his autobiography, "I wanted to cut away a big tree which I thought interfered with the view from the windows of the dining-room of our house." Although some of his family objected, he felled the tree before breakfast to avoid their interference. "So it turned out," a reflective Rockefeller concluded. Young John grew up in a world that viewed any human intervention in nature as improvement over the chaos and waste of wilderness.[11]

With the family uprooted and beginning anew in the bustling canal town of Cleveland in the early 1850s, John applied these lessons and earned a reputation as an able bookkeeper for local produce firms that connected Cleveland's agricultural hinterland to the Atlantic market. His move to the business world was made for him. His father—who had all but abandoned the Rockefeller family and assumed the name Dr. William Levingston, with a new wife in Ontario—determined John's early career arc in another way. Desiring a college education that could place him in the ministry, John received word from his vagrant father that growing family expenses would make his education impossible. As he revealed in an interview late in his life, "My father . . . conveyed an intimation that I was not to go." Instead, John dropped out of high school and enrolled in E. G. Folsom's Commercial College so as to polish his accounting skills in preparation for a job search in Cleveland. At the age when modern American teenagers take driving lessons and gain increasing freedom, John surrendered the spiritual future he had chosen for himself in order to help support his family. Between courses at Folsom's College, classmates recalled summer days spent swimming and playing tag in Lake Erie with John. In little more than a decade Rockefeller's business practices would forever alter the lake and fully realize the commercial potential of Cleveland.[12]

Having graduated from Folsom's College, Rockefeller found a position with the commodity trading house Hewitt and Tuttle. John demonstrated a talent for bookkeeping and reducing costs far beyond his age, attracting the attention of Maurice Clark—a Folsom College classmate—who proposed a partnership to connect the grain and meat of Cleveland's frontier with hungry urban populations. The partnership, brokered in 1858, would become the seed of Standard Oil's empire. At the age of eighteen John displayed the gambler's disposition of his father when he invested the entirety of his life savings in the venture. What likely would have been a modest but reliable source of income became a jackpot of riches with the start of the Civil War a year later. By mastering environmental improvements and transportation networks built in the previous generation to connect the produce of the frontier with an army of hungry stomachs, Rockefeller and Clark's business raked in no less than seventeen thousand dollars each year of the war.[13]

Few Americans understood logistics better than agricultural jobbers like John D. Rockefeller and Maurice B. Clark. The system linking productive landscapes to distant consumer markets was legible to minds able to reduce an entire river valley to a collection of goods assigned fluctuating prices. Thus, when a new resource entered this bustling market in the late summer of 1859, a mere eighty miles to the east of Cleveland, Rockefeller was prepared to incorporate it into his enterprise.

"OILDORADO"

Before the discovery of the commercial applications of petroleum, oils produced by processing animal and plant fats served a wide range of uses. Whale oil lubricated the buzzing spindles at Lowell and burned in the headlights of the first railroad engines to pierce the continent's interior.[14] As industry blossomed in the nineteenth century, the demand for illuminants and lubricants boomed. The trade eradicated cetacean populations as whaling vessels chased entire species to the brink of extinction in the Atlantic Ocean. As early as 1791 Nantucket whalers rounded Cape Horn and entered the Pacific Ocean for the first time, the Atlantic herds too diminished to profitably hunt. In 1820 the whaling fleet searched as far as the Japanese coast. By 1848 Sag Harbor whalers sailed through the Bering Strait and into the Arctic Circle in search of their prey. When Herman Melville published his romantic tale of the struggle between Captain Ahab and Moby-Dick in 1851, the whaling industry peaked as the fifth-largest American industry before entering a period of permanent decline. New England whalers lost entire crews to the gold mines of California when the vessels resupplied in Pacific ports during the 1850s. In the Civil War, Union commanders employed thirty-eight New England whaling vessels in the war effort by loading them with ballast, sailing them into Charleston Harbor, and scuttling them in the hopes of blockading the Confederate port. Markets demanded more than nature could supply, leading to the collapse of the entire industry and, more importantly, the ecological base that supported it.[15]

As fewer barrels of whale oil returned to New England ports, the value of whale oil priced it out of use for all but the rich. By 1862 oils derived from other animal fats had replaced whale oil in American lighthouses, and aspiring entrepreneurs struggled to improve the quality of lard illuminants by experimenting with lamp designs and diverse chemical combinations. Turpentine promised a sustainable alternative until the war cut off the southern timber industry from the national market. Advertisements like the one W. Lyon and Company purchased in the *Cleveland Morning Leader* in 1854, announcing "15 b[arrel]s winter bleached Whale Oil . . . Just rec[eive]d and for sale," became increasingly rare and would have caught the attention of the city's enterprising merchants.[16]

Rock oil—or petroleum—oozed from the hills of western Pennsylvania but attracted little attention. Entrepreneurs bottled the dark green liquid on their travels through the region and proscribed libations of it to frontier settlers as a cure-all for ills as diverse as cholera, bronchitis, and consumption (tuberculosis). One Pittsburgh resident, Samuel Kier, established a business to peddle "Kier's Petroleum or Rock Oil," advertising the bottles of crude oil as "Nature's Remedy Celebrated for its Wonderful Curative Powers." Kier penned a poem to market his product that reflects the hopes connected to the new commodity:

The Healthful balm, from Nature's secret spring,
The bloom of health, and life, to man will bring;
As from her depths the magic liquid flows,
To calm our sufferings, and assuage our woes.[17]

Ironically, in its early commercial life petroleum was better suited to the business acumen of William rather than John Rockefeller. It remained little more than a trifling quack medicine and a nuisance to the salt miners of western Pennsylvania until the middle of the nineteenth century.

By the mid-1850s, two New England professionals with landholdings in western Pennsylvania decided to explore the combustive qualities of the mysterious green fluid that seeped onto the surface of their property. J. G. Eveleth and George H. Bissell collected samples of the crude petroleum and paid Benjamin Silliman Jr., professor of chemistry at Yale, to run a series of experiments on the fluid and prepare a detailed report. Silliman burned the samples delivered to him in a variety of lamps, recorded the amount of smoke produced and, using a photometer, the luminosity of the flame. "I cannot refrain from expressing my satisfaction at the results of these photometric experiments," Silliman wrote in his report, delivered on April 16, 1856. "[T]hey have given the Oil of your Company a much higher value as an illuminator than I had dared to hope." By subjecting the samples to variations in temperature, Silliman discovered additional uses for petroleum. "As this oil does not gum or become acid or rancid by exposure . . . as well as . . . its wonderful resistance to extreme cold," Silliman believed his experiments demonstrated petroleum's "important qualities for a lubricator." Bissell and Eveleth could not have received better news. They secured a charter from the Connecticut legislature in November 1856 and formed the Pennsylvania Rock Oil Company to manufacture and distribute petroleum from their land in Venango County, Pennsylvania. They enticed other New England merchants to join their enterprise and reorganized the company in 1858 as the Seneca Oil Company, initially capitalized with three hundred dollars' worth of subscriptions divided into twelve thousand shares. When a salt bore run by their local agent, Edwin Drake, struck oil on August 27, 1859, petroleum stepped out of its role as curiosity and into the modern market.[18]

Within a year's time, an "Oil Rush" came to the western Pennsylvania fields. Local landholders leased their land to anyone who was able to pay inflated rates and begin drilling immediately. Lands near producing wells were gobbled up by fortune seekers and businessmen in the hopes of repeating Drake's discovery. Less than a year after the first successful strike, no fewer than twenty-one wells began operations surrounding Drake's well. As wildcat drillers sank more holes into the field, they discovered they had to drill ever deeper to extract a diminishing quantity of crude. Thus, the early, hardscrabble days of the petroleum industry were marked by misinformation, experimentation, and waste. Legal rulings reinforced this trend by applying the Rule of Capture, which declared the "owner of a tract of land acquires title to the oil and gas . . . though it may be proved that part of such oil and gas migrated from adjoining lands." The Rule of Capture evolved from case law—*Pierson v. Post* (1805) and *Acton v. Blundell* (1843)—for first animal then water resources that moved across property boundaries. Property seems like the frontier of legal claims, but petroleum presented sticky problems: What if a neighbor's well drains the oil below my claim? Do I have a right to use explosives that alter the nature of a field shared by multiple users? Although it became formally established in regards to petroleum when in 1889 the Supreme Court of Pennsylvania ruled that, "If an adjoining, or even distant, owner, drills his own land, and taps your gas, so that it comes into his well and under his control, it is no longer yours, but his," the law merely reflected decades of practice—the oil belongs to the driller who brings it to the surface.[19]

The Rule of Capture led to a mad rush into the oil fields of Western Pennsylvania. Wells that initially proved profitable slackened and went dry as a forest of competing derricks arose around them. Despite organizing a scientific, legal, and technological basis for the petroleum industry, the Seneca Oil Company was operating under a deficit by 1862. In the spring of 1864, the investors agreed to sell all assets, settle all debts, and close their enterprise. The early years of the petroleum industry did not portend its future success as a replacement for whale oil.[20]

STANDARDIZATION

Observing the birth of this new commodity from Cleveland, Rockefeller's partners approached him about entering the petroleum industry. Contrary to his image as a fearless pioneer, Rockefeller remained cautious about the new commodity. He had gambled and won by trading food in a war economy, but why press his luck on such a turbulent curiosity? "I agreed with him [Maurice Clark] that we'd go in," Rockefeller recalled, "thinking that this was a little side issue, we retaining our interest in our business as produce commission merchants." Rockefeller described the Pennsylvania oil regions as a "mining camp" run by "wild mining men." Rather than purchase producing land or try his hand at a speculative lease, Rockefeller and his partners understood that a steady supply of crude petroleum and transportation infrastructure created a situation in Cleveland that was unavailable in western Pennsylvania. By focusing on adding value to the resource by refining it into kerosene and industrial lubricants, Rockefeller could both exercise control over the resource and apply his bookkeeping skills to supply markets at low cost and maximum profit.[21]

Rockefeller, together with the Clark brothers as equity partners, entered the petroleum market by building an oil refinery at the confluence of the Cuyahoga and a tributary, Kingsbury Run, in 1863. The refinery, bordered by the river and the Atlantic and Great Western Railroad, was perfectly situated to capitalize on the region's overlapping transportation infrastructure via lake steamer, railroad engine, or canal barge. The Clark brothers brought in a fellow Englishman, Samuel Andrews, a chemical engineer who successfully refined some of the first crude petroleum into kerosene in Cleveland, to make a science of the alchemy of converting crude oil into profits. John knew almost nothing about the properties of rock oil. "Sam Andrews knew how to refine the oil," Rockefeller admitted later in life, "knew how to treat it with sulphuric acid; and Maurice Clark was desirous to get two brothers of his—to get an interest in the oil refining business." Rockefeller, however, proved a quick study and demonstrated his youthful lessons of thrift and a hatred of waste in the operation of his new refinery.[22]

Faced with, as he called it, "ruinous competition," Rockefeller fell upon the "principles of centralization" so as to capture the kerosene trade. He integrated as many factors of production as possible within the Kingsbury Run works. For example, the primary method for transporting crude petroleum were hooped wooden barrels—still the standard of measuring oil commodities despite their modern obsolescence. Rockefeller not only demanded the construction of a cooperage at the Kingsbury site but also acquired tracts of forest to supply it. Rockefeller applied his father's lessons, even allowing the freshly cut wood to dry. "We did not haul it out until it was well seasoned —when it weighed half as much as when it was green," Rockefeller recalled later in life. "This saved fifty per cent of the cartage charge." Although they questioned Rockefeller's need to extend the debts of Clark, Andrews, and Company, the Clark brothers eventually softened to the young clerk's desires. After all, with twenty refineries in Cleveland, the price of barrels soared as high as $2.50 a piece, at a time when the crude petroleum within each barrel demanded only thirty-two cents a gallon. As a commodity merchant, Rockefeller fully understood that the closer one brings consumption to production, the less waste one incurs. Like the builders of the canals and railroads, Rockefeller understood that nature's asymmetry could produce waste, and waste begot further costs. His father's woodlot lessons had fully matured in his own mind into a rational system of eliminating costs by controlling every possible stage of production, including nature itself.[23]

With the close of the Civil War, Rockefeller had invested himself wholeheartedly in the petroleum industry. In 1867 he purchased the remaining shares of the Clark brothers—who had grown weary of Rockefeller's constant desire to expand and incur debts—for $72,500. Rockefeller retained Samuel Andrews to run his refinery, which by the close of 1865 was the largest in Cleveland with a capitalization of two hundred thousand dollars. Although Rockefeller relied on Andrews's technical knowledge of refining, he grew to despise him. As an English chemist, Andrews likely rankled at Rockefeller's Baptist morals and all-consuming management of refining techniques. Later in life, Rockefeller described Andrews as a "poor workman," "bull-headed," and lacking "self-control." They would remain partners, but Andrews's role in the business would recede and his importance in the early petroleum industry all but faded into obscurity. Still five years away from the incorporation of the Standard Oil Company, Rockefeller focused on consolidating his share of the market in Cleveland. In 1865 Cleveland boasted thirty refineries—Rockefeller's the largest—with the capacity to refine two thousand barrels of crude a day. He did not wish to see the competitive chaos of Western Pennsylvania repeat itself in Cleveland.[24]

John came to believe that absolute control over management and production would deliver him from failure. When Rockefeller's coopers organized for fair wages, he locked them out and eventually replaced them with machines operated by unskilled Czech immigrants. When the economy soured in the mid-1870s, Rockefeller cut the wages of his Czech coopers from twelve to nine cents per barrel, which resulted in a general strike of all laborers making less than a dollar a day throughout Cleveland. Like the canal-builder, John straightened the crooked whims of his environment and produced barrels at a cost of ninety-six cents while his competitors purchased them at more than twice that amount. Rockefeller is best known for vertically integrating his industry and, as a result, maximizing profit by eliminating waste. His zeal for vertical integration was merely the industrial application of the same principles that had helped him capture the dry goods market. Nature was reorganized to concentrate wealth in his hands and to transfer waste or costs outside of the legal framework of his business.[25]

Rockefeller grasped the importance of measurement early in his petroleum enterprise. In 1870, the year of incorporation for the Standard Oil Company, the account books for the Lake Shore Crude Oil Transportation Company, which maintained a pipeline between Rockefeller's refinery and the train depot, recorded gallons of petroleum to the third decimal place, or about three-quarters of a teaspoon. By 1873 he employed a clerk to "send him every morning until further advised a statement of the number of barrels shipped, total gallons and average gallons per barrel." These calculations grew beyond the ability of a single man, and by the 1880s, when Rockefeller's empire grew to include the Atlantic coast, Standard Oil employed twelve men in a "Statistical Department" to keep records of the company's business and make regular reports to an executive committee chaired by Rockefeller. Ever hungry to eliminate waste, John often asked his clerks to produce cost comparison reports, such as that submitted by George H. Hopper, the general manager of Standard of New York's barrel repairing department. Hopper's report, submitted in September 1885, revealed that Standard's cooperage at Hunter's Point not only had higher throughput but was also saving $3.07 *per unit* over the company's cooperage in Williamsburg. If all of the Williamsburg production could be moved to Hunter's Point while maintaining the same cost, the company would save $2,566,695 over the course of three years.[26]

SOMEWHAT OF AN INSTITUTION

The Civil War created a hunger for news in Cleveland that newspapers would nourish as the war gave way to enterprise during the 1860s. Edwin Cowles, editor of the Republican-friendly newspaper *Cleveland Leader*, reflected the booster attitude of many in the Forest City as the Industrial Revolution took hold. Cowles marveled at the emerging wealth on display at Millionaires' Row, where Rockefeller and his associates transformed the neighborhood into "literally a street of palaces." He also took pride in a blossoming public infrastructure, bragging, "Our Water Works is the best in the West." Initially, the birth of the petroleum industry seemed to offer private wealth and civic harmony.[27]

By the end of the Civil War, Cowles wrote to his brother Samuel, the *Leader* had secured "the exclusive [right] to take the news by telegraph of the Western Associated Press." "In other words," Cowles continued, "we have the complete monopoly." Affiliated with the ascendant Republican Party at the height of its power, the *Leader* would provide Clevelanders with ample description of the industrial changes that were about to transform their urban environment in the coming decades. News media played a tangled role in the Rockefeller legacy. His harshest critics found an audience in the magazines and newspapers of the Gilded Age. Conversely, most papers of the time reflected the pro-business attitude of the national parties (especially the Republican-affiliated papers such as the *Leader*) and defended corporate capitalism from government regulation and worker grievances. In personal correspondence, Rockefeller referred to news in the *Leader* throughout his career in Cleveland. John and his brother Frank kept a watchful eye on the papers, opening one letter with "As you will notice by this morning's Cleveland Leader" before they discussed how the paper reported on events. After consolidating nearly every refinery in Cleveland under Standard's banner in the 1870s, John purchased interests in the city's two republican papers— taking out a five-thousand-dollar stake in the *Herald* and ten thousand dollars' worth of shares in the *Leader*. Decades before the advent of the public relations industry, Rockefeller understood the role of newspapers in shaping opinion, and he left nothing to chance.[28]

Although critics would accuse Rockefeller and his Standard Oil Company of buying entire legislatures, his involvement in politics was an extension of the same concern he had with the news media. Antislavery to his core, John helped a man purchase his wife out of slavery in 1859 and provided funds for the founding of what would become Spelman College (named for his wife's family) to provide higher education to the freedwomen of Atlanta. His passion for the Republican Party also issued from self-serving goals. As his business grew, John's influence extended to the state legislature in Columbus. After Standard bought out his refining interests, Oliver Hazard Payne served as a liaison between Republican legislators and JDR. During the winter of 1884, Payne reported to Rockefeller on the progress of a pipeline bill in the Ohio legislature. "It is possible that we can resort to parliamentary methods that will retard its passage," Payne wrote. Recognizing that "this act does not give any special privileges," Payne recommended against expending the company's political capital on this trifle "thereby weakening our position with the Legislature when there are matters of consequence to be considered." When legislation threatened Standard's interests, however, the company did not hesitate to apply pressure on Republicans. In the spring of 1887 the company's agents took advantage of a delay in voting, "which gave us time to spread the news throughout the state and have the benefit of our friends['] influence, which we are advised was very strong and was the means of killing the bill." Rockefeller could expect allegiance from Republican legislators. He corresponded with the Republican governor Charles W. Foster, who not only raised funds for the party but regularly purchased shares in the Standard Oil Trust throughout the 1880s.[29]

Rockefeller and Republican institutions would maintain friendly relations throughout his business career. Only during the nadir of his popularity in the 1880s did the party distance itself from the oil baron. In a letter to John in January 1887, Charles Foster admitted. "I have refused to ask you or your people for contributions for several years past" citing how "the Press of both parties of this State have been very free, as well as very wicked in their criticisms upon the Standard Oil Co." Foster, however, had overcome worries about their association. Hoping to cover the remaining twelve hundred dollars in debt from the recent political campaign, and displaying remarkable cheek, Foster wrote to Rockefeller: "My suggestion to you is, that you send me a cheque for this amount." Thirty years after entering the oil trade John was able to put his thumb on the scales of the market in his favor by buying influence over politicians.[30]

CONCLUSION

During the decade in which Rockefeller emigrated from New York and won a place among the merchants of Cleveland, the city's population exploded from 17,034 in 1850 to 43,417 by 1860. The technological alterations of the environment—the canal, river dredging, railroads, and steam propulsion—allowed the metropolis to reap the wealth of a vast hinterland that, in the case of oil, extended to Pennsylvania. Rockefeller's growing control of the petroleum industry represented a new relationship between Americans and their environment. The system of intensive agriculture brought to northeast Ohio in the early nineteenth century earned the new settlers a surplus harvest that traveled over a network of canals and railroads to feed growing urban populations. Rockefeller built the capital to enter the petroleum trade in America's frontier market economy by not only imagining the natural world as a collection of commodities in need of a market but in constructing a rationalized system to eliminate as many costs between producer and consumer as possible. He remained faithful to the simple lessons in economy that his father had taught him, and he benefited greatly from the foresight of his associates. John recognized the potential of what modern economists refer to without a hint of the macabre as "human capital." Maurice Clark and Samuel Andrews pushed Rockefeller in the petroleum business. Henry Flagler, who took John's vacant position with Clark before joining his petroleum venture in 1867, urged John in 1870 to incorporate the Standard Oil Company of Ohio. "I wish I'd had the brains to think of it," Rockefeller admitted late in life. Once positioned, however, Rockefeller's genius lay in organizing capital, workers, and nature to produce profits.[31]

His talent throughout his life also had the familiar markings of obsession. John seems to have brought his cost-cutting mentality home with him. He rewarded his children with a "wage" after performing chores such as pulling weeds or for abstaining from sweets. While many America children gain familiarity with the wage economy through an allowance, Rockefeller took these lessons to extremes, demanding that each child keep an account book and depriving them of comforts they could not afford on their own wages. With three older sisters, John Jr.'s wardrobe consisted of only dresses until the age of eight. Rockefeller seemed to extend the commodity relationship even to his friends. After building his Forest Hill estate in East Cleveland in the 1870s, John attempted to run the home as a hotel for his friends without informing them of the details ahead of time. When his guests responded to their bills with outrage and confusion, Rockefeller retreated in embarrassment.[32]

Nature, too, lost its magic in John's eyes as his business grew. He grew up tied to the land. His first responsibility in life was to milk the family's cow every morning before sunrise, which he recalled as instilling a sense of discipline during his New York years. His mother would admonish the sleepy-eyed boy, "John, if you don't wake up and set to work with a will, you'll find yourself in the county house." John remembered upstate New York as "a lovely home: healthy and beautiful country, noble views on every hand. . . . I remember the water by the shore [of Lake Owasco] was cool and clear. You could see the stones at the bottom for a long way out from shore." He also recalled catching "perch, fine big yellow ones, some of them as big as half a pound and delicious to eat." But as the demands of his oil empire encroached on his time, it became difficult to escape to such natural pleasures. "In the years when I was working hardest," he recalled as an old man, "I sometimes felt a severe pain and a sense of oppression at the back of the neck." The only treatment he sought was to "leave my office in the afternoon and drive a pair of fast horses as hard as they could go. . . . That drive did me more good than anything else." But his connection to nature became increasingly commercial as he aged. By the time he was in his seventies, and twenty years retired from the oil business, Rockefeller could not help but run his Forest Hill estate more like a business than an asylum from toil. His letters to W. B. Smith, superintendent of Forest Hill's seven hundred acres, reveal just how far Rockefeller's business ideas extended to nature:

> You might send me a list of the wood you delivered last year to our Cleveland friends. I am considering what we will do for them in this respect this year. How much wood have we on hand? At what price can you sell? Do you feel certain that we are getting as much or more for milk as our entire cost of labor, feed, etc.? What will we have for sale off the place in addition to the garden products, and what has already been sold? The idea being to see a little in the direction of determining about future agricultural operations on the place.

REFINING NATURE

To his credit, W. B. Smith not only satisfied Rockefeller's request but also reported that he had successfully sold the estate's surplus ice to a neighbor and brought in thirteen hundred bushels of oats "and a very good crop of corn." Having retired with one of the largest personal fortunes in world history, John remained obsessed with accounting for every dollar of profit he could manage to squeeze from nature.[33]

Rockefeller played a key role in alienating many Americans from the cycles of nature in the closing decades of the nineteenth century. The petroleum industry he built shattered apparent ties to the land. Where Amerindians and American subsistence farmers had calibrated their year to the cycles of food production, capitalists at the close of the nineteenth century began to transgress even the daily cycle of the sun as millions of kerosene lamps brightened the night. It is telling that, in an era when Americans increasingly associated food with urban stores rather than with farms, petroleum would cause a second sun to rise every night with little thought of the millions of years or geology required to produce a single drop of kerosene. Today we take for granted the billions of dollars' worth of infrastructure we call upon when we summon electric lighting or refrigeration with the flip of a switch. When kerosene replaced whale oil, tallow, and turpentine in lamps in the 1860s it revolutionized the way people lived their lives.[34]

A visitor to Cleveland captured these changes in a brief description of the industrializing metropolis. "The old pasture grounds of the cows of 1850 are now completely occupied by oil refineries and manufacturing establishments; and the river, which but a generation ago flowed peaceful and placid through green fields, is now almost choked with barges, tugs, and immense rafts . . . the view, though far from beautiful, is a very interesting one." John D. Rockefeller would take the lessons of his youth and apply them to Cleveland's new commercial landscape. He and his associates incorporated the Standard Oil Company in 1870, deliberately choosing a name that conveyed a sense of stability and safety in a turbulent, often dangerous new energy economy. While he shrewdly outmaneuvered his competition and won (or bought) influence in government, Rockefeller discovered that the combustible nature of petroleum frustrated his desire to control and eliminate inefficiency.[35]

TWO

In late March 1877 Peter G. Golden had an accident. While Golden was removing weights atop a boilerplate at the Standard Oil Works in Cleveland, Ohio, his supervisor ignited trays of gasoline placed below the plate without knowing Peter remained in harm's way. In a flash Golden became engulfed in flame. By the time he was removed from the fire, Golden's face, hands, and body had received terrible burns. The incident rendered Golden unable to return to his craft of boiler making and in early autumn he filed suit against the Standard Oil Company in the Court of Common Pleas for ten thousand dollars in damages. His counsel argued that Golden's supervisor had failed to notify his employees of the imminent flames and the careless use of the explosive fluid by the company rendered Standard Oil liable. The Standard Oil Company marketed itself on the safety of a "standard" illuminant in an age of experimentation, ruthless competition, and the absence of regulation. As it grew to capture 90 percent of the petroleum market by the 1880s, the company created a new combustible landscape wherever its employees and consumers refined, handled, or consumed its products.[1]

Fire historian Stephen J. Pyne has argued that humans coevolved with fire, where "each amplifies the other." Once mastered, fire served human populations by extending their power over environments. By propagating fire, humans created (at times, unknowingly) landscapes and ecosystems that became dependent on anthropogenic fire. For Pyne, however, the Industrial Revolution "rewired the dynamic of fire on earth." Once industry severed fire from its ecological role, new environmental relationships emerged that contemporary science struggled to comprehend. Petroleum historian Brian Black argues that "conflagrations were symptomatic of the overall priorities with which the oil industry viewed the entire landscape." Government, too, failed to protect communities as the antiregulatory environment of the Gilded Age encouraged a moral system based on caveat emptor.[2]

Peter Golden's fate became entangled within the logic of vertical integration championed by Standard Oil, the industrial desire to control the flow of its commodity from the derrick to the consumer. Having no market for gasoline, the Standard Oil Company sought any profitable use for the fluid. By employing a boilermaker and utilizing the dangerous vapor as a fuel for molding boilerplates, Standard at once eliminated the need to store the flammable nuisance and brought yet another level of production under its control. What happened once this flammable fluid emerged from its subterranean home, flowed through the alleyways of Gilded Age commerce, and found itself in the factories, barns, and bedsides of millions of Americans?

PRODUCTION

The Standard Oil Company's refining and distribution apparatus made oil and its products cheap and ubiquitous during the 1870s. There ensued an energy revolution in which renewable animal- and plant-based products —tallow, lard, whale oil, turpentine, vegetable oil—lost favor to cheaper petroleum derivatives such as kerosene, gasoline, Vaseline, and tar. What once took months or years to grow in forests, seas, and the range could be had with a simple salt-bore, steam engine, and a little luck. Although inexpensive, crude petroleum and its derivatives proved far more dangerous— boilermakers like Peter Golden were accustomed to using wood or coal as fuel for molding steel. While certainly also dangerous, the fuels that were replaced by petroleum burned less intensely and remained relatively immobile once aflame.

Getting at the crude petroleum deep within the western Pennsylvania earth presented its own dangers. Some pockets of crude oil, heated by the pressure and friction of the surrounding layers of rock, partially refine in situ and, once penetrated, create an array of vapor, gas, and liquid that produces the intense internal pressure termed a "gusher." Many wells, however, require active pumping of gas or water into the petroleum pocket to create pressure to draw the dark liquid to the surface. Edwin Drake employed a five-horse-power steamboat engine in Titusville, Pennsylvania, to draw his first barrel from the earth in 1859. The single-stroke engine gave a loud report every revolution, and the boiler required fresh wood to keep running. All of the equipment utilized in the infant industry had been built for some other use: the drills for boring salt; the stills, which refined the crude, were the same as those that produced whiskey; the wood barrels, which transported the crude, were picked up secondhand after holding beer, turpentine, cider, vinegar, or molasses.[3]

The combined result of this patchwork of American industries produced a baffling landscape. Barrels and holding tanks, when they leaked, soaked the soil black. Greasy rainbows swirled in the muddy puddles and wagon-wheel tracks in the streets of Titusville, Oil City, and Pithole, Pennsylvania. The scene engaged all the senses. To stand on the banks of Oil Creek in the decades after Drake's strike would require an observer to cover one's ears, as the cacophony of thousands of derricks firing in unison drowned out the sound of flowing water and the rustle of leaves. The pungent smell of oil and boiler fires burned the nostrils of those in the valley, and large natural gas flares pierced the dark Allegheny night.[4]

The wildcatters who gathered in the Oil Regions, however, pumped more than just crude oil from the ground. Not two months following his famous strike, Drake's well exploded magnificently and burned to the ground on the night of October 7, 1859. The well's engineer had carried a lantern into the engine house to inspect the works. Gas had slowly leaked from the well and saturated the surrounding atmosphere. When the gas met the open flame of the lantern, it ignited in a flash. The nephew of Captain Townshend— Drake's sponsor and representative of Connecticut investors—reported that the blast "blew everything to pieces, and started a fire," destroying the pump and derrick. The explosion ruined the head start Drake had secured in the nascent industry, and he watched the hills around his little well grow loud with the sound of steam engine pumps as he rebuilt the works. Three years later the Seneca Oil Company ran into debt. On March 7, 1864, the company Drake helped found went under, not five years since he lifted the first bucket of crude from the Pennsylvania soil.[5]

In the decade following Drake's strike, fires and explosions became such a serious threat in western Pennsylvania that some well owners posted signs that cautioned "Smokers Will Be Shot." Despite such efforts, the underlying fact remained—bringing oil to the surface required fire in the bellies of nearby engines and lanterns. As the rest of the nation mulled the repercussions of the coming Civil War, five days after the fall of Fort Sumter, a spectacular strike near Titusville resulted in one of the first true "gushers." At twilight the city emptied as residents made their way to witness the spectacle of a fountain of oil reaching sixty feet into the sky, a black lake forming at its base. As the astonished crowd circled the dark geyser, it ignited from an unknown source and exploded with violent force. As the black fountain transformed into a pillar of flame, burning oil splashed the crowd of spectators, who struggled in vain to extinguish their oil-soaked clothing. Ten people died and eleven more were severely burned. The fire raged for three days until workers smothered it with a mixture of soil and manure.[6]

STORAGE

While the gusher fire of 1861 ignited from an unknown source, the particular flammable properties of petroleum render it prone to ignition from any source that raised it to its burn point, the temperature at which it combusts. When pumped from a well, crude petroleum at the molecular level looks like a bowl of spaghetti. The braided hydrocarbon necklaces that form the mass of crude oil, although stable, can alight with a little heating and direct contact with a flame. Nature fixed an indirect relationship between the length of a hydrocarbon chain and its flammability. While asphalt contains thirty-five carbon atoms (or more), its cousin propane has only three. By applying heat or pressure, crude petroleum begins to separate into its constituent molecules. If you apply enough of either, you can "crack" the bonds between the atoms and create "lighter" and less stable hydrocarbon molecules. Twentieth-century industry safety texts list a panoply of phenomena to avoid in the presence of petroleum—lit cigarettes, hot boilers, friction, lightning, and static electricity. Even spontaneous combustion can produce misfortune. Oily rags, left in a pile and cutoff from circulating air will undergo spontaneous heating as the carbon fibers slowly oxidize. If enough time transpires without disturbance, the temperature will eventually rise to the oil's flashpoint and a mound of discarded rags will become a ball of flame.[7]

Storing crude oil proved just as dangerous as drawing it to the surface. In the first decade of the industry, most storage tanks for crude were made of wood and placed in close proximity to one another. They leaked incessantly and were prone to spectacular accidents. In 1866 such an accident occurred on the Bennehoff Run in the oil regions of western Pennsylvania. During a thunderstorm, lightning struck a metal pipe and ignited a nearby storage tank, which exploded and sent a torrent of burning crude into the stream. As the burning slick made its way down the run, it set fire to and destroyed no less than twenty derricks and storage tanks. At the bottom of the run, the wave of fire finally surrounded a three-thousand-barrel storage tank. One of the only tanks made of iron, it survived the flames and spurred producers to replace wood with hardier materials.[8]

The new iron tanks, when they did catch fire, however, required inventive techniques to extinguish. The jumble of technology in the nascent petroleum industry appropriated the tools of war to fight the menace of fire. With the Civil War just concluded, former soldiers-turned-wildcatters heard the familiar report of cannon whenever an iron tank caught fire. Cannonballs fired into the metal tanks proved handy in letting out the excess oil of a storage tank that would otherwise burn for days. From these experiences it was clear that, even in its most stable form, petroleum carried inherent natural dangers that required patience and experience to master.[9]

Brian Black argues, "Americans were fully willing to write off" fire losses in Pennsylvania's producing regions "if it could provide them a steady supply of valuable crude." Economic imperatives created a new geography of industrial sacrifice zones, landscapes surrendered to fire, pollution, and dramatic transformation. In the spring of 1885, the secretary and future vice president of Standard Oil William P. Thompson wrote to John D. Rockefeller about problems with the company's storage facilities in Cleveland. The City Council of Cleveland found that a tank yard owned by the company on Todd Street violated a council ordinance prohibiting any storage vessel "built within 200 feet of a house or street." Although facing a suit for damages, Thompson advised Rockefeller to allow the construction of the tank yard to continue. "There is no greater damage likely to be awarded for 4 more tanks than our present number," Thompson reasoned. Confident that Standard could outmaneuver the city council, Thompson declared the construction would succeed "even if we have to surround that property with a brick wall." Standard Oil and local governments would continue to negotiate the legal landscape for storing volatile fluids. Two months after alerting Rockefeller to the legal shortcuts to the company's storage problems, Thompson wrote to inform him that Standard's main warehouse in New Orleans burned to the ground, with damage estimates of around fifteen thousand dollars. Few other industries feared their commodities could endanger people and property while resting in storage.[10]

TRANSPORTATION

The transportation of petroleum presented a host of dangers for the young industry. Historian Christopher Jones has argued that the petroleum energy transition would not have been possible without creating a "landscape of intensification" that controlled the pesky "fluid nature of petroleum." Production of a cylindrical railroad tank car for petroleum, now familiar to modern eyes, did not emerge from the hodgepodge of extant technologies until 1869. To move crude oil from the regions to the refining centers of New York, Pittsburgh, and Cleveland meant either loading it in several dozen forty-two-gallon wood barrels or in a few open-air wooden tubs perched on railroad flatcars. Both methods proved costly—the former required a large quantity of hand-crafted barrels, made scarce by the oil boom. Both leaked fluid and gas and were prone to loss from evaporation.[11]

Before the penetration of railroads into Pennsylvania's oil regions, the cheapest method of transporting crude oil was via scow or barge down the Allegheny River to Pittsburgh. Although the river remained shallow throughout most the year, shippers established a complex organic machinery to confront the inadequacies of nature. In 1862 the producers appointed a superintendent of shipping who paid the owners of local sawmills to coordinate the release of freshets from their milldams and collected apportioned fees from the shippers who rode their barrel-laden barges down the river on these human-induced floods. As one mode of production helped usher in the next in the Pennsylvania backcountry, the proximity of human and flame remained close.[12]

During a spring freshet in 1863 a lantern overturned on a barge filled with un-barreled bulk crude and transformed the vessel into a floating pyre. The flames spread from boat to boat in the choked river until nearly a hundred craft carrying no less than eight thousand barrels burst aflame—along with a bridge that spanned the Allegheny. Carrying oil by railroad car was hardly safer. After a particularly costly fire, the general freight agent of the Pennsylvania Railroad stated, "I would rather carry anything else than oil in tanks." And with good reason: an early railroad fire was caused when several tank cars crashed and released burning oil into a nearby sewer where it flowed, still aflame, into and overflowed a canal. The blaze resulted in five hundred thousand dollars in damage to surrounding property. Oil, the freight agent concluded, was "worse than powder to carry."[13]

In unloading crude at transshipment points and refineries, laborers confronted many of the same often unseen dangers that were present at the well and storage tank. On a warm July morning in 1873, for example, a shipment of Pennsylvania crude arrived at the Standard Oil Works on Forest Street in Cleveland. The morning sun lay below the horizon, and Charles McFarland, Robert McDowell, and Chris Osterland set about unloading the tank cars by lantern light. The oil, sealed in the iron tank car for the entirety of its trip, released gas when the men attempted to transfer it to Standard's storage tanks.

Upon meeting the flames of their lanterns, the gas ignited. A Cleveland police sergeant rushed to the scene after hearing the explosion, but the men were beyond aid. The three men "had been baptized by fire" and "roasted alive," according to a *Cleveland Plain Dealer* reporter. Ignorance of the invisible dangers associated with petroleum combined with a careless landscape permitted such accidents to revisit the property and bodies of American oil producers for decades following Drake's strike.[14]

Standard Oil earned some of their most significant competitive advantages in transportation. By increasing the throughput of his refineries in Cleveland, Rockefeller won rebates from railroad companies that were desperate for the security of a high-volume client. Railroad baron Jay Gould proposed the rebate system so as to secure dependable freight on his Pennsylvania line, but Rockefeller and Flagler soon wielded it to their advantage with every rail carrier. Railroad companies became so desperate for Standard's business that they paid "drawbacks"—sometimes as high as forty cents—to Rockefeller on every barrel of a competitor's oil shipped on the line. The nation's railroads, spurred by deep corruption as well as by a sincere government desire to extend economic and political control over the West, overextended in the thirty years following the Civil War. By the depression of 1893 many railroad companies had failed or been consolidated into other lines.[15]

Fully enjoying this buyers' market, Rockefeller also pressed his advantage by manipulating the liquid nature of petroleum. By 1885 Standard controlled the Continental Oil Company (later Conoco) and its fleet of tank cars, which maintained the illusion of independence from Rockefeller. In December 1887 Frank Rockefeller wrote to John, revealing Standard's less celebrated business practices. Frank forwarded to his brother data on thirty-one tank cars in use by Standard (via Continental), including "two sheets, one giving the actual capacities, and numbers of the cars; the other giving the capacities as furnished by Mr. M.A. Robinson to the railroads." Standard was underreporting the volume of its tank cars so as to save on shipping. "The difference," Frank reported, "is 15,643 gallons, equivalent to 504.61 gallons per car or 10.09 barrels per car." The purpose of Frank's letter was not to gloat but, rather, to inform John that, "owing to the extreme necessity for tank cars during the past few months[,] some of these cars have been run in the general business and not confined to the Continental Oil Co." This exposed the lie of Continental's independence and "would place us in a very bad light" if discovered. Frank acknowledged that the tank cars currently in production would only be "62 gallons per car short of the actual capacity." A more modest "difference that some of our people seem to think could be easily explained away before the Interstate Commerce Commission." Because the Lake Shore Railroad "are gauging and weighing cars," Frank asked his brother if "we should furnish the railroad with the actual gauge or furnish them with a gauge representing 62 gallons short of the actual capacity." Similar to the case of Peter Golden and the tank yard on Todd Street, Standard's practices in transportation reveal a drive to press the legal boundaries of Gilded Age commerce by manipulating the nature of petroleum.[16]

REFINING

The process of converting crude petroleum into marketable products was similar to that of distilling alcohol. All one needed was a metal tank capped with a gooseneck nozzle, some piping, and a good fire. As the fire heated the crude oil, the more volatile, or "lighter," hydrocarbons vaporized, rose to the top of the still and escaped out the gooseneck. The vapor then descended a long, usually spiraled, pipe called a "worm" that was cooled with water. The lowered temperatures of the worm condensed the vapor into a distillate, which then dripped into a tank awaiting below. Different grades of oils distilled out at varying temperatures, but a primary distillation yielded an average of 20 percent gasoline, 70 percent kerosene, and 10 percent wax, grease, coke, asphalt, and other heavy residuum. The distinction between the grades was illusory, however. Light kerosene, that is kerosene with lower flash- and fire points, is no different from heavy gasoline. The words represented a human invention for a vague range of properties where nature made no distinction.[17]

It was easy for anybody with the ability to boil water and with access to equipment to start up a backyard refinery in the infant industry. However, refining capacity, throughput, and a zeal for expansion and low unit costs determined who could exercise power over competitors and control the industry. After the incorporation of the Standard Oil Company in 1870 and its acquisition of every large refining interest in Cleveland over the following two years, the corporation controlled the destiny of nearly every barrel of oil that made its way through Cleveland. Standard's refineries required proximity to the Cuyahoga watershed for both manufacturing purposes and access to markets. The refineries and tank yards crowded the Kingsbury and Walworth runs that emptied into the Cuyahoga, drawing water off to distill kerosene that would leave by rail or float downstream via steamer to Lake Erie and points east. Cleveland's refining capacity skyrocketed during this period. While the entire country produced but two thousand barrels of *crude* oil in the year following Drake's strike, the mayor of Cleveland boasted that his city alone refined and sold 1.5 million barrels of petroleum in 1871.[18]

As owners bought out competitors and managers ensured low labor costs, all united against the common enemy: fire. Rockefeller, disgusted with waste, became obsessed with containing fire to the boilers. "I was always ready, night and day, for a fire alarm from the direction of our works," Rockefeller recalled. When the inevitable accident occurred, as it did four times between Thanksgiving and Christmas in 1873, Rockefeller "dashed madly to the scene of the action" and, if finding the blaze beyond control, "would have my pencil out, making plans for the rebuilding of our works."[19]

Standard maintained its own private firefighting equipment but also relied heavily on the municipally funded Cleveland Fire Department for all but the smallest blazes. Early in Cleveland's history, the citizens of Cleveland deemed a firefighting force too essential to its survival to be left up to the whims of the market and apportioned city taxes to its maintenance. Private and volunteer fire departments had proved a disaster for public health. Other cities that experimented with several companies haphazardly rushing through the streets to arrive first at a fire, found that accidents and even fistfights or gun battles among the competitors proved as deadly as the flames the crews were rushing to extinguish. The public fire department was a nineteenth-century marvel, merging martial organization with the latest technologies. As early as 1829 a volunteer firefighting force patroled Cleveland, and in 1863 the city council authorized a force of fifty-three men organized by a chief engineer and paid by and responsible to the public. At the close of the Civil War, Cleveland's fire chief could boast a public infrastructure of five steam fire engines that could draw on 167 hydrants supplied by 50 reservoirs, all at an annual cost of $63,904.11. Mayor H. M. Chapin reserved a section of his annual message to heap praise on the department. "Our steam Fire Department is justly the pride of our citizens, and for efficiency and good management, is probably not excelled in the world," he beamed. Although the department "is expensive, and adds to our taxes," the mayor assured a frugal city council "the decreased cost of insurance because of their use, annually returns to tax payers more than threefold the cost of sustaining this system." By 1879 the fire department maintained 161 fire alarm boxes, installed by Western Union, including a new box at the Standard Oil Works on Kingsbury Run. Understanding the public interest in the city's response, the fire department alerted and often picked up reporters from the *Leader* when answering an alarm.[20]

Necessity guided the development of a public infrastructure as the business of converting Pennsylvania crude on the banks of the Cuyahoga led to epic fires, requiring the services of the entire municipal firefighting force. The telegraphic fire alarms used to summon the force were remarkably sim-

REFINING NATURE

ple. At the outbreak of fire all one had to do was swing open a protective panel, grab hold of a large knob, and pull it down a grooved track. As the knob slowly made its way back to its original position, an automated telegraph transmitted the number of the box to the fire department telegraph officer who then alerted the appropriate fire engine crew.[21]

At three in the afternoon, on Monday, February 23, 1880, the busy box at the main Standard works transmitted an alarm after the bottom of a still suddenly gave out, releasing three thousand barrels of crude, which "ignited and spread in all directions." To increase throughput and efficiency the refinery was packed with dozens of stills, which quickly became engulfed in flames as the fire spread from one still to the next, growing in size as each still successively exploded and contributed its contents to the conflagration. As workers fled the wave of fire, the burning oil spread to two further Standard distilleries and, following the grade of the land, flowed into the Kingsbury Run, "which ignited, and the current threatened to carry destruction to other works lower down the run." A reporter from the *Cleveland Leader* marveled at the efforts of the firefighters who dumped streams of water on the flames and, failing to halt the burning river, "engaged in building dams to stay the progress of the water." Although the Standard Oil Company met with "the worst conflagrations it has ever experienced," the efforts of the fire department saved the bulk of the works and contained the burning slick to a small stretch of the Kingsbury Run. Standard Oil was confronting a literal firewall. The more Rockefeller packed his stills and tanks closer together to squeeze ever more production from the same real estate, the more he invited the threat of calamitous fires.[22]

Fire, eventually, entered into Standard's cost analysis. In a letter to Rockefeller in 1883, secretary William P. Thompson proposed "diminishing" the night watch at the company's Cleveland refineries. Thompson's logic was simple: "by reducing this force, if a fire was to start, the probabilities are the men disposed of would not be of very material advantage in stopping it, and if it did burn we would get the pay for it." Thompson recognized the opportunity to transfer Standard's costs to a third party—the insurance agency— assuming "satisfactory rates can be obtained." The executive committee of Standard Oil balked initially, despite Thompson's pleas that "in the event of a fire, the whole business would be swept out and the loss entailed a very large one." Thompson's concerns issued from the company's experience with refinery fires throughout the 1870s and 1880s. While some of these fires took advantage of careless plant design or greedy operation, some of the largest capitalized on Standard's environmental illiteracy.[23]

THE INFERNAL REGIONS

Late winter was always a precarious time for the lower Cuyahoga. Snow melt, a saturated water table, and frequent storms flooded the natural flood-plain, or "flats," of the Cuyahoga. The commerce of the Gilded Age, however, shared the same landscape. Mills, breweries, and oil refineries choked the flats and used the river as a source of water, a commercial highway, and a sewer. Unlike the river, the infrastructure of capital proved immobile. On February 4, 1883, the hydraulic system of northeast Ohio reclaimed some of its lost real estate. The front page of the *Cleveland Leader* stated simply, "The rain fell, and the ice thawed, and the floods come." Cleveland awoke to a horrifying, altered landscape. Lake Erie appeared to have disobeyed its boundaries and invaded the Cuyahoga's course, turning the industrial corridor of the flats into "an immense lake," "dimpled as a laughing baby" with hundreds of whirlpools and eddies. The Kingsbury Run ballooned to twenty times its normal size and engulfed a leaky still of the Standard Oil Works. In its flow, it carried petroleum (in all its various states) downstream with it. This freight of mud, water, and oil reached the refinery of the Great Western Oil Works downstream. A cannonlike report shattered the morning air as Standard's oil, loosed by the river, met the boiler fires of the Great Western Works and exploded at six in the morning. Fire crews rushed to the scene, but they found their efforts frustrated by rushing water. "There was no use trying to put the fire out, so it was allowed to burn until the oil was exhausted," a writer for the *Cleveland Leader* reported.[24]

The firefighters busied themselves with extinguishing the burning oil slick on the swelling river, which eventually entered the Standard Works proper, forcing them to rush back upstream. As leaking oil ignited yet again and made its way downstream from the Standard Works, firefighters raced the burning slick downstream, where a wooden track belonging to the New York, Pennsylvania and Ohio Railroad lay in its path. The firefighters attempted to raise a floodgate and halt the burning river, but the force of the water proved too much, and the men established themselves at the base of the track with the goal of preventing its destruction. As the noon hour passed, stills and storage tanks at the Standard Works exploded, releasing their oil into the flood and turning the sky black with smoke. The occasional explosions, whirling smoke, and burning river led one *Leader* reporter to biblical imagery. The "combination of noises, together with the roaring of the fire and the weird shadows thrown upon the surrounding streets and buildings, completed what might justly be termed a true picture of the infernal regions."[25]

Thousands of people, lured by the sound of exploding tanks and the rising column of smoke, crowded around the perimeter of the works. The pyrotechnics created by the commingling of petroleum, water, and fire entertained and threatened the crowd, who had taken to the tops of rail cars for a better view. When the intense heat finally buckled and melted the exterior of a large storage tank, water rushed into its belly and carried away a wave of burning oil, causing the crowd to flee in terror. As the day wore on, Standard Oil and generous citizens donated coffee, food, and tobacco to the firefighters. By 9:30 that night, the massive explosion of a sixteen-thousand-barrel tank caused residents living in earshot to throw their valuables into horse carts, ready to flee the flames.[26]

From the first explosion at the works, fire crews gathered the rubble from the destroyed yards and pitched them into the run to corral the flames. By three in the morning, the extent of destruction was of such magnitude that the run was nearly plugged. The crews worked throughout the morning and into the next afternoon. Two firefighters had to leave the scene after sustaining injuries. The Standard Works, in the end, was completely destroyed, but the efforts of the municipal force contained the damage to a relatively small area on the Kingsbury Run. No less than sixty thousand barrels of petroleum—crude, kerosene, and gasoline—fueled the two-day fire. At least two businesses benefited from the inferno—the streetcar companies found their tracks flooded with spectators and a saloon adjacent to the works "was running full blast . . . three bar-tenders being kept busy dealing out drinks and poor cigars" to the throng of onlookers.[27]

It was a miracle that the blaze spared the greater part of the city, a fear on the minds of many in the wake of the great fires of Chicago and Boston of the previous decade. Despite the hazards created in bringing petroleum from the ground to the banks of the Cuyahoga, the river bound its flammable consequences to the industrial corridor. Once refined, packaged, and sent to retailers, however, kerosene left the concentrated productive landscape and spread across the globe and was insinuated into every darkened room that a human wished to illuminate.

THE NEW ALCHEMY

On their way to Illinois from New York in the early 1870s, women's rights advocates Elizabeth Cady Stanton and Susan B. Anthony passed through Cleveland in the middle of the night. They woke from their slumber as the illuminated city filled the windows of their railroad car. Rockefeller's success in expanding the market and distribution of kerosene transformed the urban night and startled observers. The consumption of kerosene had grown from just under a million barrels at the time Rockefeller purchased his first refinery in 1863 to nearly seven million barrels in 1873, when the Standard Oil Company refined a plurality of world kerosene. By 1880 the company refined 90 percent of all American crude petroleum. Stanton and Anthony, stunned by the glowing city, questioned the train's conductor who informed them that the inexpensive fuel in Cleveland permitted women to work long into the night to accomplish the unfinished tasks of the day. From its advent, the rising of this second sun proved far more problematic than boosters claimed.[28]

In her largely autobiographical novel, *Quench the Lamp*, Alice Taylor recalls her childhood in rural Innishannon, Ireland, during the 1940s before electricity had made its way to farming communities and replaced kerosene. Her memoir suggests that, even under the best conditions, the use of a paraffin oil lamp required constant maintenance and care to avoid injury:

> Earlier in the day the globe of the lamp had been washed with lukewarm soapy water and polished with a soft cloth or newspaper. The base had been filled with paraffin oil and the wick had been trimmed. All was done in readiness for the night because leaving these jobs till the natural light had gone could lead to breakages in the dusk, to overflowing oil and to frayed tempers as a consequence. . . . At first the wick was kept down low after lighting to give the globe a chance to warm slowly, because sudden heat could crack it. A hairpin was sometimes hung on top of the globe to prevent this happening.

Taylor's memoir confirmed Anthony's observation—that the lighting technology reinforced domestic responsibility for women.[29]

Although kerosene clearly brightened Gilded Age cities, stunning observers such as Anthony and Stanton, its use did not represent a complete revolution in technology. Just as the drilling and refining stages applied preexisting technologies to the new industry, consumers burned kerosene in lamps that had changed little over the centuries. Oil lamps had existed since ancient times. Any permeable material that could hold vegetable oil or animal fat could light a room when floated on the surface of the reservoir and lit. At the close of the eighteenth century, Genovese inventor François Pierre Ami Argand had advanced this design by separating the wick from the reservoir and surrounding the lighted wick with a glass cylinder to improve luminosity by regulating the oxygen draught over the flame. This new design increased luminosity and ensured stable combustion but also required the regular cleaning of the lamp chimney and introduced more movable, and therefore, breakable parts.[30]

Kerosene's price advantage over renewable illuminants made it a natural choice for most consumers—two and a half gallons of kerosene provided the same luminosity as thirty-seven pounds of sperm-oil candles and did so at one-seventeenth of the cost. Standard's business strategy left nothing to chance in the company's drive for market domination. From its incorporation in 1870, Standard controlled distribution and marketing interests in Cleveland and East Coast markets. By the mid-1880s, Standard held marketing firms in every major market, and its 313 bulk stations by 1906 had grown to 3,573. Many wholesalers signed exclusive contracts, agreeing to carry only Standard products in exchange for a guaranteed profit, so wholesalers could undercut Standard's competitors and grow their market share. Like the railroads before them, wholesalers found themselves guaranteed of steady sales but often chafed at the prices dictated to them from Standard's headquarters.[31]

Standard offered consumers varying brands of illuminating oils, such as Pratt's Astral, Royal Daylight, and Peerless. These brands not only communicated expectations of an illuminant's quality, they also masked Standard's near complete dominance of the kerosene market. As Standard purchased competing refineries, the company shrewdly retained the former trademarks so as to ensure consumer continuity and to quell whispers of monopoly power. Internally, Standard officials reduced illuminants to two categories. Water White oils passed burn tests of 150°F, contained no impurities, and burned clean with little smoke or scent. Common oils, on the other hand, simply passed the government's 110°F test and offered consumers an adequate illuminant at a lower price. At the height of its control of the illuminant market in the 1880s and 1890s, Standard struggled to supply the semi-independent sales wing of its empire with enough illuminants to meet demand.[32]

Balancing these spinning plates proved beyond the means of Standard's executives by the 1880s, and problems emerged that threatened the company's control of the industry. To squeeze out competitors, in certain markets Standard would dictate price cuts that frustrated wholesalers. By the late 1870s Standard recognized that the price wars had taken a toll on their marketing agents. H. A. Hutchins, who would retire as head of the Domestic Trade Department, warned Oliver Payne in the spring of 1879 that a recent reduction of a half cent on the price of their illuminants in the upstate New York market strained relations with retailers. The price cut, which targeted a competing refinery in Titusville, when communicated to the New York retailers "of course made them dissatisfied," Hutchins reported. In another letter, Hutchins reveals that steep competition forced a retailer in Cincinnati to sell at a loss. Hutchins offered to reimburse the marketing company on the condition that they "send us a report at the end of each week of the number of barrels of O.S.T. oil sold."[33] Proud of their sales abilities, Standard's retail agents often took poor sales figures personally. Forced to undercut competition in New York City, G. B. Burton reported a loss of $2,600 on sales of Water White oil. Burton wrote to the executive committee, whose representative George Gregory explained to Rockefeller, "He feels that your Committee should not consider this apparent deficiency to be the result of a lack of good management on his part, and I think they should not, for he has managed his large territory with ability and success." Some retailers chose to betray the executive committee's price manipulation rather than suffer losses. W. P. Thompson, a Standard executive, discovered that the P. C. Hanford Oil Company in Chicago "let some trade slip rather than make losses" by undercutting competition. After "quite a struggle," Thompson forced Hanford to slash prices for its Water White oils, advising Rockefeller that the situation "would be better, if we had a stronger and more capable representative with them." The economic war with competitors strained the relationship between Standard and its retailers. Both confronted material limits as stocks of the higher quality Water White oil dried up and would remain scarce throughout the 1880s.[34]

By the late 1880s Standard executives confronted a supply problem that would plague the company throughout the kerosene age. In a report to the executive committee in the summer of 1888, F. Q. Barstow, secretary of the manufacturing committee, informed the company, "we shall be short 400,000 bbls. of a sufficient quantity to meet the wants of the trade." In a letter to the Domestic Trade Committee, Barstow suggested, "a greater reduction than that already authorized should be made in the price of common oils" in the hopes of drawing some consumers away from their Water White stock. Despite their efforts, sales figures for Water White grew by 20 percent that year while Standard's total domestic sales increased by only 12 percent, weighed down by decreases in common oil sales. The price wars of the 1880s no doubt drove the cost of Water White into the budget of more consumers, but the rising specter of lamp fires may have also motivated Americans to pay a premium on safer oils. The flammable properties of petroleum once again served as a limit to Standard's control.[35]

As the company continued its price war with competitors, the internal correspondence of Standard Oil turned to concerns about the quality of their illuminants. Standard agent Feargus B. Squire alerted John D. Rockefeller in the summer of 1889 that a run of twenty-five thousand gallons of Water White flashed at 109°F, suggesting it would fail to meet the standard fire test of 150°F. Although Standard's refiners had "pumped it back into agitator and are spraying and blowing it, trying to get test up, which I know will be a hard job," the failure suggests that Standard struggled to maintain the quality of its commodities. Rockefeller had already voiced his concern to F. Q. Barstow in 1883. "I do not wish to complain," Rockefeller began, "but the great embarrassment at Cleveland on account of the poor quality of the oil has not been less serious than we have estimated and I hope that if necessary to help out that extraordinary efforts will be made on the part of New York to hurry shipments. What has been the cause of the delay?" A year later, when Standard's Water White failed tests conducted in Minnesota, Rockefeller directed Standard's treasurer, Oliver Payne, to investigate the Cleveland refineries for the source of the problem. Payne reported that such a complaint, "upon investigation, was regarded as not very well founded." Payne argued that the refiners operated on "the most correct principles of refining and were producing the most satisfactory oils ever made in Cleveland." Payne appears to have dismissed the Minnesota test. "In the immense business that we are doing," Payne wrote, "it is almost impossible not to have a few complaints."[36]

John's brother did not share Payne's sanguine disposition. Frank Rockefeller believed the company's refiners concealed faulty practices out of pride. "As manufacturers," Frank wrote in a letter to John, "we are naturally very anxious ourselves to protect the good name of the various brands of oil turned out by us." After receiving an anonymous report of failed tests from Kansas City, Frank decided to conduct his own research. After a battery of tests in June 1889 Frank reported he had found "in a number of cases the fire test is considerably below 150, and in some instances below 140." Frank believed these figures demonstrated "that compounding has been done with lower grades of oil than 150," which explained why Water White 150 degree oil accounted for "99 per cent of our complaints." His findings also opened a wedge in an already rocky relationship with his brother that would eventually lead to their estrangement. "This is one of the reasons why I have always insisted with you," Frank admonished his brother, "that it was absolutely necessary for the selling Committees to have representatives continuously in the field making personal inspection of all of our own and associate interests." Playing at alchemy with oil quality was the art of refining, but it also contained a lure to forsake quality for profits.[37]

Since the founding of Standard Oil, John D. Rockefeller had aimed to produce oil *at*, not above test. This may explain Rockefeller's contempt for proud chemists such as Sam Andrews. Early correspondence between the highest-ranking members of Standard Oil revealed that tests "of our oil" had shown "burning about 113 . . . 110[°F]" (ellipsis in original). The letter illustrates John's desire to squeeze as much money as possible from his stock without falling below test. The correspondence reported, "John think can [illegible] cut down a little," adding "but be careful and not cut it down too much."[38] By the 1880s, however, the price war had pinched Standard's retailers and drained stocks of its quality Water White illuminant. Had sales jobbers stretched their Water White supply by mixing in common oils in order to save face and secure profits? Did refiners, under heavy pressure from the executive committee to satisfy consumer demand, water their stock with lower quality illuminants? The imperatives of market control threatened to undermine Standard's hallmark: safety and quality.

As Standard struggled to fill lamps, it sought a market for its more volatile gasoline stocks. Gasoline, a by-product of refining and roughly 20 percent of a refined barrel of crude, proved worse than dead weight. Sitting in storage tanks at refineries, the fluid loomed like a time bomb. The properties of gasoline frustrated Rockefeller, who sought any use for "this lighter product, the naphtha, gasoline." Rockefeller's experiments "to burn it for fuel in distilling the oil" proved failures. Gasoline proved too explosive, as Peter Golden knew all too well. It also seemed to cost more than alternatives to use as a fuel. Rockefeller ran a five-year experiment at his Cleveland refineries to ascertain the cost of using his own oils as fuel for heating boilers. When completed in 1890, the report found that coal maintained a price advantage, saving the refinery as much as thirty-two thousand dollars a year. For years, Rockefeller admitted, his refineries simply dumped gasoline in the nearest waterway. "[A]nd thousands and hundreds of thousands of barrels of it floated down the creeks and rivers, and the ground was saturated with it, in the constant effort to get rid of it."[39]

Standard eventually organized a "Gasoline Stove Department," headed by marketing experts who released fifteen hundred advertising agents into the American market to extol the virtues of gas stoves to American housewives. They mounted a full-scale modern advertising effort by distributing pamphlets, and they flooded newspapers, trolley cars, and magazines with advertisements. Despite this quarter-million-dollar ad campaign, demand for gasoline failed to strip refineries of the product until the turn of the twentieth century, and the rise of the internal combustion engine. Ironically, Rockefeller faced a dilemma when tenants of his real estate properties in Cleveland installed the very stoves his company hoped would transmute gasoline into profits. In 1888 his personal real estate agent, J. G. W. Cowles, discovered that "two of our tenants in rooms in the Woodland avenue block (Mr. Case and Mrs. Mie.) use gasoline stoves for cooking." This should have come as little surprise, considering that Standard also used a property in the same building for "display of stoves for sale." The problem, however, issued from an "additional cost of insurance in consequence of the use of gasoline." Cowles suggested, because the tenants did not have leases, "we can require them to leave if we wish, or make it a condition of their remaining, that they stop using gasoline." So much for the illusions of Standard's stove department. "I do not see," Cowles concluded, "how our stove store can be restricted in this matter," which simply paid the increased cost of insurance in rent. Whether due to the insurance burden or the fire hazard, Rockefeller did not want his tenants operating stoves that he was actively promoting to consumers.[40]

ACCIDENTAL LANDSCAPE

In the thirty-year window between the incorporation of Standard Oil Company and the advent of electricity, the company rose to dominate the petroleum illuminant industry almost completely on the sale of kerosene. While the price advantage of kerosene captured nearly all the attention of consumers and commentators, its chemical properties put it at a significant disadvantage in one significant respect—an overturned tallow candle or whale oil lamp, while dangerous, lacked the explosive nature of kerosene. A kerosene lamp, if upset, could easily douse the surrounding area with burning kerosene, whereas only the wicks of animal and plant-based lamps and candles remained lit upon a spill.

Nineteenth-century urban landscapes proved ripe for large conflagrations, even in the absence of kerosene. During a dry and windy autumn in 1871 Chicago had burned in an epic fire that ended in the destruction of two hundred million dollars' worth in property damages and left one-third of the population (one hundred thousand) homeless. The beginning of the 1870s revealed many of the environmental limits Americans approached. The Great Chicago Fire along with a firestorm in the timberlands of Peshtigo, Wisconsin, which spawned a tornado of fire, led commentators to dub 1871 the Black Year. A year later, a fire raged through Boston, destroying 776 buildings in the process.[41]

Fire and its causes captured public attention just as kerosene emerged as the predominant lamp illuminant. Mirroring the era's laissez-faire attitude in economics, Cleveland Fire Department chiefs believed fire, like disease, emerged in an atmosphere of loose morals. The fire commission in Cleveland believed the wooden hovels of the poor immigrants represented a fire menace and the "equivalent to putting on kindling preparatory to a large fire." The editors of the *Cleveland Leader* laid blame for the Chicago fire on a "bummer government . . . made up of Germans, Irish and the whisky drinking portion of the native born population, [whose] costly steam engines might as well have been in the hands of schoolboys." The editorial broad-side—entitled "Paying the Piper"—ended on a grim note stating, if "the majority of Chicago voters prefer free rum and Sunday revelry to efficient government and a well managed fire department, of course it is their funeral, not ours." Only a small but increasingly vocal segment of the population looked to the new lamp fuel as a source of the combustible urban landscape. The majority of the press and city officials relied on familiar scapegoats by highlighting racial and class stereotypes instead of confronting structural fire regimes exposed to increasingly combustible technology. "So long as cities continued to rebuild themselves on the same pattern and with similar materials," environmental historian Stephen Pyne has argued, "they experienced the same kinds of fire."[42]

Cleveland fire chiefs documented the increasing occurrence of kerosene lamp fires in their annual reports. The City of Cleveland suffered only 56 fires in 1865, before the ubiquity of kerosene fueled lamps (although two of those fires started at refineries). Nine years later, in 1874, the city's department responded to 249 fires, of which 11 came from exploding oil, 5 from spontaneous combustion, and 13 from exploding kerosene lamps. These numbers no doubt underrepresent actual figures, for many fires began in the absence of a human witness. In this same year, fire crews responded to 100 fires of undetermined cause. What is surprising is that not a single fire was attributed to either candles or whale oil lamps, suggesting that kerosene represented a novel consumer hazard. Even children playing with matches caused more fires than either of the standard illuminants of the pre-kerosene era.[43]

The extreme volatility of kerosene produced sudden, dangerous blazes. Three weeks before the *Leader* editors laid the cause of Chicago's misfortune on drunk and careless firefighters, a servant girl at Cleveland's City Hotel, performing her night duties by kerosene lamp, burned herself severely when she accidentally upset the vessel. Of course, lamps did not always require jostling or overturning to cause disaster. As the midsummer sun set on Cleveland in 1872, Mr. and Mrs. Hofshild lit a kerosene lamp. A short while later the lamp exploded, showering the couple with burning kerosene. Mr. Hofshild's sister extinguished the flames, but not until the couple and their rescuer had been severely burned and one thousand dollars' worth of property destroyed.[44]

Contemporary observers of the petroleum industry grew aware of the explosive threat of kerosene lamps. British petroleum scholars Captain J. H. Thomson and Boverton Redwood recorded 129 fatal lamp accidents, per year, in England during the 1880s and 1890s. The Society of Chemical Industry of Great Britain put the figure at "about three hundred deaths a year" for England and Wales. Thomson and Redwood recorded the many variables that gave rise to these numbers. "[S]mall glass lamps which are sold for a penny, and on which the burner is fitted without proper attachment" easily spilled their contents when upset. These and other lamps could become incendiary bombs if the vapor content of their reservoirs inched up to 2–4 percent, breaking the brittle glass and showering the surroundings area in burning fluid.[45]

Of course, if discriminating consumers chose a strong lamp they could still become victims if the kerosene they purchased to fill it had been made "light" by the addition of gasoline. In February 1867, the US Congress established a fire test of 110° Fahrenheit for all petroleum illuminants—requiring all kerosene to burn only at temperatures above 110° Fahrenheit. Five years later the General Assembly of the State of Ohio adopted the 110° test, making manufacturers and retailers liable for damages to property and persons. Yet, the state law appropriated no funds for inspecting illuminating oils and capped fines at a maximum of one thousand dollars or twenty days of jail time. Thus, the sale of below test illuminants had few disincentives. The problem became exacerbated because, in passing the 1867 ordinance, Congress adopted a "fire" and not a "flash" test. The difference is important. Illuminants that burn at 110° Fahrenheit can form ignitable vapors, or "flash," at temperatures much lower. In a study endorsed by the Board of Health of the city of New York following the passage of the law, Dr. C. F. Chandler found that oil passing a fire test of 110° F often flashed, or formed explosive vapors, at 70° F, or the temperature of a warm spring day. The findings led Chandler to judge the federal fire test "as worthless" and recommend a more appropriate flash test of no lower than 100° F.[46]

In the early months of 1872, the city council of Cleveland appointed a "Non-explosive Committee" to investigate the city's illuminating oils following the rise in fires resulting from kerosene lamps. After the committee tested various illuminating oils from vendors throughout the city, they found "*not one was up to the legal standard of fire test.*" The results led the usually pro-business editors of the *Cleveland Leader* to urge regulation in their columns: "some legislation is imperatively demanded for the protection of the community against these dangerous illuminating fluids."[47] By 1875 knowledge that manufacturers and retailers regularly adulterated their kerosene with more volatile fluids had become widespread. The *Cleveland Leader* reported "that peddlers are buying coal oil of proper test and mixing benzine [*sic*] or naptha with it, thus making a spurious and dangerous article," and linked the "many lamp explosions recently reported . . . to this poor oil." Petroleum, which required a trained and patient hand even in its most innocuous form had been thrown into an explosive environment—an unregulated market that rewarded cost-cutting and a popular culture that eschewed corporate responsibility with the attitude of caveat emptor.[48]

By 1876, four years after the Ohio legislature and nearly a decade since Congress established the 110° burn test, the continued rise in fires associated with exploding lamp oil (twenty-two that year in Cleveland alone) and the failure to implement an inspection mechanism led citizens to take up the mantle of inspection in the public interest. S. Buckland and his son wrote in to the *Cleveland Leader* after testing "a large number of samples sold at retail in this city" and found that "it flashes below 100, and ignites below 110." At a meeting of insurance companies in Glasgow, Scotland, tests run on Scottish, American, and Russian kerosene revealed a flash point of 69° F for the American oil, confirming European beliefs that American self-interest was turning European lamps into time bombs. Even the Cleveland fire department, which always sought to blame the human closest to the source of the blaze, admitted that "some" of the twenty-two fires caused by exploding lamps in 1876 "were undoubtedly from the use of an inferior grade of oil." Yet, they repeated the refrain "'People cannot be too careful with fire' . . . an injunction which many are apt to forget." With such low flash and fire points, illuminating oil became a hidden peril as temperatures rose in summer and rickety lamps leaked explosive vapors.[49]

Even friction and static became dangerous catalysts. Cleveland resident Cora B. Gage received severe burns to her hands and arms while cleaning her family's picture frames with benzene. The fluid, raised to its flash point by the friction of cleaning, "burst into flames and ignited her clothing." Thomas Fela, a worker at a Cleveland steel mill, received a similar surprise as a "coal oil can he was filling from a barrel exploded" without apparent cause, "burn[ing] his arms, hands, and back." A fire at the Standard Oil works in late 1873 illustrated the pervasive danger of a world filled with petroleum illuminants. Investigators traced the blaze back to a box of "oiled rags and sweepings from the floor" that had spontaneously combusted without cause. The writers at the *Cleveland Leader* reported the incident with little interest as it was "pretty well established that this thing is possible, and this is not the first fire in Cleveland which has been traced to this cause."[50]

REGULATION

In Victorian Great Britain and France, where governments were more active in taking on the onus of protecting public health, the advent of the oil age elicited immediate response. In France strict guidelines regulated the storage of kerosene to cool, ventilated lots at least forty-five meters from residential areas. On June 29, 1864, only five years after Drake's strike and the beginning of the kerosene age, the Conseil d'Hygiène et de Salubrité publique de la Seine created and distributed to the public a pamphlet, approved by the police commissioner, on how to handle and maintain kerosene and lamps safely. Two years later the pamphlet's recommendations were enacted into law, structuring the entire landscape of petroleum production on the Continent—in stark contrast to the United States.[51]

In the United States, despite regulatory laws, a popular knowledge of the risk of petroleum illuminants, and the outcry of citizens, federal and local governments failed to regulate the petroleum industry. After a rash of lamp fires and several deaths, Cleveland fire chief James Hill urged, "Some measures ought to be taken to prohibit the sale of oil which is not up to the standard . . . to better protect the lives and property of our citizens and their families." Despite a "general impression . . . in the community . . . that the annual revenues of the city were insufficient to carry on its government properly," the chief auditor of the city reported in 1874, a debt of "$1,822,000 paying an annual interest of $127,540." Despite the wide acceptance of the public to provide funds for fire security, the city governments of Cleveland in the closing decades of the nineteenth century refused to appoint a city inspector of petroleum illuminants to enforce the federal and state lamp oil laws.[52]

Cleveland's mayors during the Gilded Age viewed the city's responsibilities for public oversight of business as costly and, at worst, a threat to the growth of enterprise. In defending a low 15 percent tax rate (10 percent of which supported a fire department), Cleveland mayor Charles A. Otis argued for the most rigid economy in his final annual message to the city in 1874. His successor, N. P. Payne, echoed this fiscal conservatism in his inaugural address, calling for a "retrenchment in expenditures rather than . . . any addition to the tax levy."[53] Serving as mayor of Cleveland during the turbulent year following the great labor riots of 1877, R. R. Herrick used the platform of his own inaugural address to elevate the laissez-faire ideals that gave the Gilded Age its name. Although the "rapid growth of this city during the past fifteen years, has called for large expenditures for improvements in order to keep pace with the increase of population," he argued, "[t]axation is burdensome and to a certain degree oppressive . . . at this time." "Nothing," Herrick concluded, "so effectually strangles manufacturing and all business enterprises as a high rate of taxation."[54] Herrick put his faith in an unregulated economy that would gain traction in the Gilded Age and lead to a nationwide race to the bottom, as state after state sought to lower taxes and gut regulations in order to attract business: "Capitalists in seeking a location in which to invest their money in business enterprises look as much to the rate of the tax levy at that place as to any other one thing, and will shun a city where the rate of taxation is excessive. It should therefore be our aim, kept constantly in view, to cut down and keep down the rate of taxation in this city as much as is consistent with its best interests." That year the municipal Cleveland fire department responded to 333 fires, a nearly fivefold increase over the number fifteen years previous at the close of the Civil War.[55]

Lawmakers guided by Gilded Age notions of government oversight created state inspection teams; however, such guardians of the commonweal received remuneration from the companies they inspected, not the public they were charged with protecting. By the late 1880s Standard Oil and its affiliates paid oil inspectors as much as skilled tradesmen, their wages falling between those of carpenters and coopers. Although placing the cost of regulation on the corporations that necessitated it can be seen as a way of forcing companies to internalize the cost of their actions, the absence of intermediaries between inspectors and their corporate paymasters created a dangerous conflict of interest. Although tax-fearing city officials applauded the free-market regulation, others vested with protecting the public welfare suspected quid pro quo. In 1888, after a particularly combustive year in Boston where a tenth of all fires resulted from exploding kerosene, the fire marshal collected samples of oil and sent them to chemists at the Massachusetts Institute of Technology. While the state oil inspectors declared similar samples above standard, the professors found all the samples sent to them to be below the state test. In 1890 a deputy inspector in Iowa wrote a formal charge to the governor against his fellow lamp oil inspectors, accusing them of abandoning their posts and leaving their official state stencils (used to certify oil) with the various companies. The governor demanded that the inspector produce witnesses, which killed the investigation before it began, and sealed the documents relating to the case from public scrutiny until he left office. A year later the Committee on Illuminating Oils of the Minnesota Senate affirmed testimony against its state inspector, which charged that he let oil company employees use his official stencil to brand kerosene as they saw fit. In his testimony, the inspector defended his actions with the simple declaration, "I am under no obligation to the State of Minnesota. The Standard Oil Company paid me."[56]

In the event a suit was brought against the company following an explosion, Standard's lawyers relied on the absence of enforcement to create a fog of uncertainty as to the quality of their own oil. In a 1905 case involving a Michigan man, the company's legal counsel argued that "no recovery could be had in the absence of a showing that the defendant had actual knowledge of the inferior quality of the oil." Such knowledge was impossible to establish, Standard's lawyer argued, because the "oil inspector was extremely lax in his methods and his inspection fell far short of compliance with the statute." Fortunately for the plaintiff, the victim of terrible burns when Standard's adulterated oil exploded, the presiding judge held the company liable for the damages sustained to his person and property.[57]

After reviewing the trade in 1885, US Department of the Interior agent Stephen Farnum Peckham recommended the creation of a special committee to test and oversee the sale of petroleum products and enforce a more vigorous test with stiff penalties. Peckham believed that consumers would only benefit with regulations "making the selling of dangerous oil a misdemeanor in all cases, and manslaughter when death is occassioned [sic] by its use . . . and also providing for the recovery of damages for all losses to either persons or property." No such committee emerged in the antiregulatory environment of the kerosene age, which passed without a centralized apparatus to audit the manufacture and sale of illuminants in any meaningful way. Despite a kaleidoscope of local and state laws intended to ensure public safety and consumer protection, the Standard Oil Company enjoyed extraordinary freedom from government oversight as the market spread its perilous commodity to every corner of the world.[58]

CONCLUSION

Three years after receiving debilitating burns, the boilermaker Peter Golden accepted an out of court settlement from the Standard Oil Company. After the company argued that Golden "had neglected to follow directions" and "[a]ll injuries sustained by him were due to his own carelessness," the testimony of a coworker and witness, Frank Shulte, sank any chance of having the case dismissed. Shulte testified that Golden's supervisor, Jacob Hummel, "without looking, set fire to the gasoline." We will never know how much the company offered Golden to avoid an official judgment, but in settling out of court, the company preserved their status as an unaccountable party. Just as the words "careless" and "accident" obscured the history of lamp fire victims, the settlement shrouded the story of Peter Golden.[59]

At the Paris Exposition of 1867, a gas engine patented by the Belgium-born engineer Jean Joseph Étienne Lenoir, captured the attention of the audience by safely harnessing the explosive force of petroleum. As internal combustion engines transitioned from a mere curiosity to a legitimate source of motive power, the Standard Oil Company finally found a market for the nuisance fluid gasoline, just as coal-powered electric power stations made their dangerous kerosene lamps obsolete in urban areas around the world. Kerosene lamp oil had already helped birth a new mode of production into existence—the modern era would forever be tied to the fate of inexpensive, often dangerous nonrenewable resources. John D. Rockefeller succeeded in his goal of "building up as rapidly and as broadly as possible the volume of consumption" of petroleum illuminants the world over. In 1870, the year Standard incorporated, the United States exported 97.9 million gallons of kerosene. Fueled by the aggressive marketing and distribution of the Standard Trust, that figure skyrocketed to 740 million gallons at the apex of the kerosene age in 1900.[60]

The explosion of the trade had a dark side in the growth of public costs. In 1868, when Rockefeller ordered his refiners to bring his oil down to the test standard set by Congress, Cleveland experienced 149 fires, of which the fire department attributed 4 to lamp explosions. In 1896, after decades of price wars that transformed Standard into a monopoly power, Cleveland's fire department responded to 1,000 fires. Cleveland's fire chief reported that 124 of those fires resulted from the use of gasoline or kerosene, including 26 cases of exploding lamps. Standard Oil's market was much larger than Cleveland. As its products infiltrated new markets in the last decades of the nineteenth century, exploding kerosene lamps resulted in as many as 6,000 deaths a year.[61]

THREE

WATER

On a February day in 1874, Cleveland's city council gathered at City Hall to inaugurate the recently completed waterworks tunnel below Lake Erie. The council dressed themselves in old ragged clothes and waterproof coats for an inspection tour through the subterranean artery they had authorized and built. A *Leader* reporter took note of the humor in their curious act, remarking, "they looked more like a lot of bummers bent on having a spree than a dignified body of city officials." After a brief carriage ride to the shore station, the strangely attired crew slipped, shuffled, and fell over each other as a frigid rain fell. These men were not accustomed to wearing work clothes.

Once inside the station, the men took seats in "small cars used for transporting the dirt excavated from the tunnel" and were conveyed by way of mule train through the cave's length of one and one-quarter miles. With only the small lanterns they carried to light their way, the gang sang "Down in the Coal Mine" in a bid to lighten their spirits. At the end of the track, the men disembarked from their cars, forced to walk bent over at the waist in the four-foot tube in the final stretch to the lake intake crib where a tug awaited to carry them home. The tunnel proved "distressingly small to these of ample proportions." One council member gave up and "sat down in a foot of water, and nothing but a graphic description of the horrors of remaining there over Sunday could shake his determination not to take a step further." Once up a crude ladder to the lake platform and safe aboard their waiting tug, the crew erupted in all manner of cheers, proclamations, and song. One "prominent citizen officiated at the whistle and blew long and loud, in utter defiance of the whistle ordinance," once they reached the safety of the Cuyahoga. Their journey under the lake traced, in reverse, the new pathway of the city's water supply.[1]

Having inspected the waterworks, the council could declare to their constituents that the high cost and prolonged wait would soon pay off by delivering clean water to their taps. The old intake source, located only three hundred feet from the lakeshore, could now close and with its passing so too—the council hoped—would cease the complaints of the entire city concerning the quality of the water they drank daily. Viewed optimistically, technology had delivered Cleveland from the mercy of its environment. However, it was industry, and not blind forces such as population growth, that created a water quality crisis beginning in 1860. When the Cuyahoga captured the nation's attention in the late 1960s, a century of choices made by industry and local government contributed to the catastrophe of the burning river. The decision to refine Cleveland's water system enabled its city government to avoid regulating industry while also offering constituents a plan for the future. The water crises associated with the petroleum industry today have their origins in the nineteenth century. When oil enters water supplies, we rightfully blame industry for advancing private profits over public welfare. When the problem becomes permanent, it is apparent that public officials have failed to defend the communities they serve. The environmental history of water in Cleveland illustrates a collective— at times, collaborative—failure by both business and government.

From the end of the Civil War until the turn of the century, Cleveland's debt skyrocketed as the city government constructed ever more ambitious water facilities. The technological system the city built to refine its polluted water supply represented social as well as mechanical engineering. Despite community unrest, public officials continuously chose engineering schemes instead of containing pollution at its source—the factories, refineries, and slaughterhouses that lined the Cuyahoga. Thus, each culverted stream, gravity sewer, and waterworks pumping station represented a public expenditure so as to avoid private wastes. Populations often confront natural limits that require technological systems to supply the necessities of life, but here there were many cases in which engineering works stood in as first resort solutions to industrial problems. A progressive narrative has described the creation of such urban infrastructure as a rational response to unprecedented urban growth and its attendant thirst for an expanding water supply. Much of Cleveland's waterworks were not built to supply the city with water. They were built to supply the city with water that was free of acid and petroleum.

PIPE DREAMS

Dirty water existed before the Industrial Revolution even in Cleveland. The Cuyahoga River's mouth, choked with sandbars and timber snags, smelled of rotting vegetable matter and bred swarms of mosquitoes in the summer. The Erie Indians preferred the clear water upstream, before the swift flowing river became a slow, at times stagnant, pool. American settlers met their water needs by tapping ample subsurface water with wells and by venturing to the lakeshore for pails of clean lake water. During the decade of the 1840s when the city's population nearly tripled from 6,071 to 17,034, the maze of subterranean water wells, privy vaults, and cesspools threatened the public health during a decade of cholera and typhoid outbreaks. The lake offered a simple solution at a time when other cities (such as New York) required elaborate systems to tap far-flung water sources. Drawing water from Lake Erie required little more than the laying of pipe beneath its surface, the construction of a pumping station to provide the necessary pressure to draw it to shore, and a maze of pipes and reservoirs to distribute it to the public. After two years of labor, in 1856 a fifty-inch pipe laid eight hundred feet northwest of the Cuyahoga's mouth went online, supplying the burgeoning metropolis with clean lake water.[2]

Like many American rivers at this time, the Cuyahoga served as a crucible of industry. The river attracted breweries, tanneries, rolling mills, and slaughterhouses because of the convenience of water transport, and flowing waters carried effluent to the lake. Residential waste also entered the river via runoff from streets and tributaries before the creation of waste treatment and sewer facilities. Despite the use of the river as a general sewer, people continued to bathe in its waters, and the new hydraulic system, with its intake pipe a short distance from the Cuyahoga's mouth, remained clean. Only with the advent of petroleum refineries along the river's banks did the troubles really begin.

Crude oil flowed into Cleveland from western Pennsylvania, and kerosene, lubricants, and asphaltum flowed out. Before refiners could convert the multicolored sludge into commodities, the distillation of petroleum required a wide array of inputs to cleanse it of impurities. Sand, dirt, salt, sulfur, and anything else buried with it could create uneven combustion, foul odors, or a smoky burn. In the first significant publication detailing the refining process, the chemist Abraham Gesner described how crude petroleum enters a tank called an "agitator" at a refinery, where various "charges" of sulfuric acid, soda ash, and lime would be mixed in with the crude oil. After each cycle, freshwater would be let into the tank and agitated into a caustic soup, where the spent acids would settle to the bottom of the agitator to be siphoned off—usually into public waterways. If the crude oil contained a high level of sulfur, an alkali solution of lead oxide would also make its way in and out of the agitator. No standard formula guided the quantities used; only the acquired knowledge of the refiner determined when a batch was ready for the still and distillation.[3]

By the middle of the twentieth century, once many of the inefficiencies of the refining process had been overcome, one barrel of oil required between ten and fifteen pounds of sulfuric acid to cleanse, or about two hundred gallons for every one thousand barrels of crude. By 1872, Standard Oil commanded a *daily* refining capacity of ten thousand barrels, producing as much as two thousand gallons of sulfuric acid, a figure supported by the company receipts of the nearby Grasselli Chemical Company. Constructed in 1866 next door to Rockefeller's primary refinery on the Kingsbury Run, the Grasselli Acid Works shipped on average just over two thousand carboys of sulphuric acid a month to the firm of Rockefeller, Andrews and Flagler during the winter of 1868–1869. While the records fail to provide the size of the carboys, they typically run from a standard size of five and up to fifteen gallons, easily matching the ratios recorded in the twentieth century. The acid, once it fulfilled its purpose, was dumped directly into the river. For Standard Oil, the Cuyahoga served as both a source of pure water to cleanse their crude and as a sink for wastewater and acid. "In early days," Rockefeller recalled later in life, "after the oil had been cleansed with sulphuric acid there was a residue which was allowed to run to waste."[4]

Rockefeller selected the sites of his refineries with water in mind. His primary refinery on the Kingsbury Run just south of downtown benefited not only from the flowing water of the run but also from gravity-fed springs. Soon after establishing the refinery, Rockefeller purchased the land of David Short, which stood on the bluffs overlooking the Flats. Short's land contained ample freshwater springs and had housed Dr. Thomas T. Seelye's "water cure" clinic and spa in 1848. Several of Cleveland's prominent citizens took advantage of the "daily baths, wrappings and massage" that made up

Dr. Seelye's water cure, which promised to cure everything from dyspepsia to "nervous invalidism." Although Rockefeller took an interest in the water cure himself, he purchased the land in 1873 so the Standard Oil Company could commandeer the water rights to the property.[5]

TROUBLE IN THE WATER

After a decade of supplying the city with pure lake water, the Cleveland waterworks became the subject of numerous newspaper articles and public discussion. In 1866 the drinking water of the city darkened and tasted strongly of petroleum. The city council hired J. Lang Cassels, a professor of chemistry at Cleveland Medical College to analyze the city's drinking water. The subsequent report found the city's water compromised, which spurred public anger. The editors of the *Cleveland Leader* turned attention away from the refining interests and focused it instead on the city government. In an editorial response to a modest call by the *Cleveland Herald* demanding that the oil refiners burn their refuse rather than dump it into the river, the *Leader* editor "object[s] to this most decidedly," declaring "There is nothing unhealthy about the refuse benzene." Bringing attention to the other possible sources of pestilential organic pollution from slaughterhouses and brewers, the editor pleaded, "By all means let the oil refineries alone."[6]

As the year 1866 drew to a close, with the city's water still contaminated with the runoff of Cleveland's many refineries, the editor of the *Leader* suggested a solution to appease all parties. The city of Chicago—regularly mocked as a pedestrian, drunken circus—had solved its water needs by boring a large tunnel under Lake Michigan. The metropolis shared many affinities with Cleveland. It also sat on the shore of a large lake. The Chicago River, like the Cuyahoga, "is the great central sewer of the city, where all the offal and filth . . . is discharged immediately into the lake." The editor of the Republican *Leader* viewed the tunnel from the perspective of temperance, calling the engineering works a "sensation" in freeing Chicago from dirty lakeshore water that might induce the inhabitants to replace their "rye juice" with clean water.[7]

A year passed in Cleveland with little action. By November 1867, the *Leader* declared the "matter is becoming a serious one," pushing the city to act by announcing, "It is high time that the city government examine into this matter." In fact, the city government was well occupied. As far back as March 1866, a month after the initial complaints of the petroleum taste, Superintendent Joseph Singer had delivered a special report to the trustees of the Cleveland waterworks. Singer discovered that lake ice plugged the river's normal discharge into the lake during winter months, causing it to flow westward toward the water intake crib. In concluding, he offered two possible solutions to the problem. The city could either, through "the strictest enforcement of the existing city ordinances, [defend] against the pollution of any stream or water-course tributary to the Cuyahoga river, and [do so by] compelling the slaughter houses, oil refineries . . . and other factories, which use these streams now as a sewer, to dispose of their waste in some other way." Or they could sacrifice the river's health as a fait accompli and spend "about $500,000.00 for the construction of a tunnel under the lake, similar to that in progress of construction in Chicago." Singer argued for the former solution, citing the ease of drafting a strong ordinance and the desire to avoid the "necessity of doubling the indebtedness of the city" through the construction of such a massive system.[8]

The trustees, wasting no time, wrote the mayor the next day echoing Singer's recommendations. The city council, on their advice, passed an ordinance prohibiting petroleum refiners from dumping refuse into the watercourse under the threat of a fine. Amazingly, the ordinance excepted "spent acid and alkali" from control, ensuring that the problem would continue.[9]

In the context of American legal philosophy, these actions appear less strange. English common law had a strong tradition of protecting landholders along a waterway from injury caused by upstream neighbors. The legal historian Morton J. Horwitz explained the origins of this anti-developmental interpretation of the law in early America: "The premise underlying the law as stated was that land was . . . a private estate to be enjoyed for its own sake. . . . [I]nterferences with the natural flow of water, including both diversion and obstruction, were illegal 'without the consent of all who have an interest in it.'" The relatively unexploited North American environment helped erode the anti-developmental tenor of English common law in the American context. An 1810 case argued before the New York Supreme Court saw this mind-set fully employed when the court defended a landowner's exploitation of resources by arguing, "what would in *England* be waste, is not always so here."[10] Since their state-granted charters recognized American corporations as public servants, courts often ruled "development" of a watercourse as "beneficial to the public" despite the increasingly private nature of corporations in the late nineteenth century. Beginning with the advent of waterpower in New England, American courts chipped away at the common law protections of downstream and upstream stakeholders so that by the end of the Civil War, Horwitz argues, "a gradual acceptance of the idea that the ownership of property implies above all the right to develop that property for business purposes" became American legal orthodoxy.

Aggrieved Clevelanders who might turn to nuisance law found similar protections for capital.[11]

Nuisances—including manufacturing waste, noise, or the general disruption of the landscape—became signs of economic activity and were protected under the developmental orthodoxy of American law. By the Gilded Age, as Horwitz points out, "the essential attribute of property ownership was the power to develop one's property regardless of the injurious consequences to others." The sanctity given to private property by the political revolutions of the eighteenth century, according to legal scholar James Willard Hurst, created a new legal order in the nineteenth century that would "protect and promote the release of individual creative energy to the greatest extent possible." Thus, the courts mirrored the stance of public officials in Cleveland who equated the success of corporations with that of the public and who in most cases refused to interfere with development. Party politics played an insignificant role in an era when both democrats and republicans promoted urban prosperity by managing public resources while asking little of business. Of the seventeen mayoral terms between 1863 and 1900, democrats held office seven times, republicans nine, and an independent served once. Not until the election of the reform-minded Tom Johnson in 1901 did the office serve as a platform for significant changes in the relationship between business and the public. By then, American legal cases had transformed private property from a revolutionary political idea that protected an individual's civil liberty to a conservative economic bulwark against public control over resources.[12]

THE DEATH OF LOVELY LURLINE

The 1860s witnessed the birth of the petroleum industry in the Forest City. Lacking a single refinery in 1860, the petroleum industry then ranked as the largest industry by value of products in 1870 according to the records of the Cleveland Chamber of Commerce. With a value of just over 4 million dollars, the petroleum industry accounted for just over 15 percent of all capital in Cleveland. As John D. Rockefeller captured nearly the entire refining capacity of the city in the early 1870s, warning signs washed up on the lakeshore. In November 1867, a year after the council's ordinance, the owners and operators of the steam and sail vessels utilizing the port of Cleveland presented an extraordinary petition to the city council, which stated that the acidic waste in the river not only stripped the paint from the ships but also disintegrated the oakum caulk used to seal wooden hulls, which led to dangerous leaks. The petition urged the city government to bar refineries from dumping their waste acid into the river and to create a special committee to investigate the claim and suggest possible remedies. The *Leader* responded by confirming the putridity of the "infernal Cuyahoga water" but focused attention on their preferred remedy of an engineering fix. Here was the ultimate "Revenge Effect" of industrial development: the by-products of enterprise threatened one of the foundations of the city's wealth.[13]

The *Cleveland Leader* mounted a defense of the petroleum trade that would make the editor of the *Wall Street Journal* blush. After documenting the "immense number of dead horses, cattle, dogs, cats, rats, the thousands of tons of offal of our slaughter houses and packing houses," the editor of the *Leader* described the city's water supply as "consisting mainly of the excrement of a population of 80,000." Compared to this "delectable mess," the *Leader* claimed the "refuse of the coal oil refineries is not, properly speaking, *filthy*." The editor feared abating refinery waste would prove fruitless and steered readers to the promise of technology. "Instead of taking steps to have the inlet pipe removed from the filthy neighborhood of the mouth of the Cuyahoga to a point, say, three miles west, or to have a tunnel similar to that of Chicago," the *Leader* concluded, "[the city council members] are simply trying to remove *the indicator* of the presence of filth, by ordering the coal oil refineries not to empty their refuse matter into the river, and thus keep our people in blissful ignorance of the fact that they are drinking the most horrible filth." Their concern about organic pollutants should not be dismissed during a century of urban cholera epidemics; yet, the caustic petroleum waste, which was literally eating away at the hulls of ships, was for the *Leader* a benign chemical canary in the waterworks system.[14]

For nearly a year the committee contemplated a remedy that would satisfy the community, the shipping interests, and the refining industry. In the late winter of 1868, a Clevelander expressed the helplessness of the situation in "Song of the Sick Water-Nymph," a poem that appeared in the *Leader*. In it, the river siren Lurline—who appeared in contemporary poems and an opera—finds her river water replaced with petroleum and animal wastes as an impotent city government fails to protect the public good:

Faugh! What a smell!
How can I be well?
Stinking again,
Small pipe and main,
Even large reservoir
Yields to its power! . . .

Petroleum, slaughter-house gore,
To say nothing of acid
Sulphuric, and how many more
In our waters placid. . . .

Resolutions, reports,
"Close your retorts!"
"Our acids are spent."
On money intent—
Runs the stench to the river,
Or rises to heaven forever!

. . . . The oil that in garish days shimmers
By gaslight ever glimmers.
If sweet water from outermost sea
By chance in the summer day cheer you and me,
We remember it only as what has been.
For shortly again
It will be all azuline,
The death of lovely Lurline.[15]

A municipal committee organized to investigate the nuisance of the Walworth Run, the location of a Standard Oil refinery, reported that water pollution posed an additional risk to the city for "water thus poisoned also runs through an ice pond, from which hundreds of loads of ice are annually taken and dispensed to customers throughout the city." Once released from factories, waterborne waste entered residential water taps or ice chests in unexpected ways.[16]

In April 1868, the council ordered John Vial, the chief engineer and superintendent of Cleveland's waterworks, to meet his counterpart in Chicago and inspect their new tunnel. Having succeeded Joseph Singer, who had favored abatement over technology, Vial conformed to the probusiness politics of the Gilded Age. After assessing Chicago's tunnel, Vial announced it a success and estimated that Cleveland could build a similar works at a hefty sum of $296,000. Although he warned that Chicago's new two-mile tunnel furnished water that "is not at all time clear," the council moved, without further study, to put on the ballot a loan for the purpose of constructing a similar tunnel. By September 1868, ten months after the initial petition from the city's shipping interests, a mass citizen assembly met at the board of trade and left with a petition in response to the council's ordinance. In it, they noted the additional fire hazard of petroleum waste upon the water surface and called on Mayor Stephen Buhrer "to take immediate and efficient measures by which to suppress the further deposit, the refuse, or dregs, or any inflammable matter from the petroleum refineries." The city council ignored such concerns, preferring to abandon the river as a wasteland and focus on extending the waterworks beyond the reach of industrial pollution. At the beginning of 1869, Mayor Stephen Buhrer proudly announced the passage of the issue by public vote.[17]

During the nearly five years required to build the tunnel, complaints from the public continued to mount. Not wishing to impede business, the Republican *Leader* attempted to quell the public's passion for a more stringent ordinance by defending the refiner's claim that they "would be obliged to suspend operations entirely unless allowed to permit this [acidic] water to escape." Despite having echoed the public's disgust at the state of the city's drinking water, the *Leader* reasoned "it is certainly better that an oily taste should occasionally get into the drinking water than that any course should be taken which would seriously impede the oil interest of the city." For those waiting for the completion of the lake tunnel, they suggested in a front-page editorial the "best course for the people to pursue is probably to 'grin and bear it,' and endure."[18]

The final cost of the tunnel is lost, but subsequent governments found themselves straddled with ever-increasing debt. In 1870, when the tunnel excavations had just begun, the city's debt stood at $2,009,000. By the time of the tunnel's completion five years later, it had ballooned to $5,160,000. By 1874 Mayor Nathan Payne admitted the debt had become an "embarrassment" and "urge[d] greater economy in expenditures." It proved an ill-fated time for the city to plunge into debt.[19]

Following the economic depression in Europe, the United States entered what became known as The Panic of 1873. Banks closed, nearly one-quarter of the country's railroads went bankrupt, and the unemployment rate soared to 14 percent. Businesses survived by reducing labor costs through wage cuts or by laying off workers. John D. Rockefeller retained his labor force in Cleveland but cut costs by employing them only half-time. When the price of oil plunged to new lows in the spring of 1877, he responded by cutting the piecework wage of his coopers from twenty cents to ten cents per barrel. Standard Oil provoked a walkout by the coopers when later that year the company announced a further wage cut from ten to nine cents per barrel. The coopers, consisting primarily of Czech immigrants, demanded a resumption of their already slashed wages, which amounted to less than a dollar a day.[20]

REFINING NATURE

Before the advent of refrigeration, ice companies captured river and lake water in ponds such as this one on Cleveland's waterfront. Sanitation officers feared the industry could serve as a vector for disease transmission.

Credit: Cleveland Public Library/Photograph Collection, CP03269.

When Rockefeller and Standard's management refused to negotiate, the coopers began a months-long general strike by laborers in the city of Cleveland making less than a dollar a day. Carpenters, masons, and rolling mill laborers, "dropp[ed] their tools and refus[ed] to work in various parts of the city, showing that the feeling is contagious." This "revolt against their employers," as the Republican *Leader* termed the strike, eventually succeeded in raising the labor standards for every group but the coopers. Rockefeller refused to concede, and when the coopers' strike eventually broke, Standard Oil forced a further humiliating concession on them—they could no longer bring their sons to labor with them, a source of additional family income and a dying vestige of an apprentice craft economy. The workers had failed, in the words of their organizer František Skarda, to "equalize the differences between labor and capital" through "self-government of trades . . . that they be not dependent on their employers, but shall be their own employers."[21]

Although Standard Oil avoided the violence that spread throughout the nation during the summer of 1877, which historians identify as the Great Railroad Strike, the labor tumult of the late 1870s and the economic downturn had captured the city's attention at a critical moment for the health of Cleveland. The creation of the new tunnel was seen as the panacea for the city's social as well as environmental woes. With the perennial return of the lake ice, the population devised a number of schemes to clear the river mouth of the obstruction in the hopes of regaining clean drinking water. One *Leader* article described the use of powder explosives deposited into drilled holes in the ice as a solution. An inventive Clevelander suggested he "fit up a sled with a boiler and a steam engine, connected with two circular saws, six feet in diameter, placed six feet apart" to serve as an ice-cutting machine. One fed-up citizen believed all-out war against the ice necessary and even called for the use of the city's artillery. In the end, nature frustrated the political goals of the city council, who believed they could engineer freshwater and avoid prohibitive legislation against the petroleum industry.[22]

Only one officer patrolled the river to enforce the already toothless dumping ordinance. The refineries easily circumvented even this attempt at enforcement. When the patroling officer was recalled to police court for five days in July 1875, the citizens watched as oil was "allowed to run from the refineries into the river." The *Leader* editor in July offered an apologia by stating, "the oil refineries grew careless" during the officer's "absence." Wanting what they saw as effective government, the workers of Cleveland took the issue into their own hands. When a new refinery was proposed for the seventh ward of the city in August 1875, a confederation of laborers petitioned and won the support of the board of health in opposing its construction, arguing it, as the *Cleveland Leader* put it on August 18, "would endanger the health of the people." A week later, they delivered their petition to the city council and effectively blocked the refinery's construction. In the midst of a financial crisis, Clevelanders decided the environmental costs of a refinery outweighed its economic benefits.[23]

FLUSH AND FORGET

During the violent summer of 1877, the *Leader* and the Cleveland city government escalated their rhetoric, supporting what they viewed as capitalism under siege. The *Leader*, which had at first sympathized with the Czech coopers in their strike against the Standard Oil Company, now defended business. The editors regarded labor strikes as "defiance of the rights of property, a violation of the plainest, most fundamental provision of human government," adding "The country cannot be ruined that [laborers] may gain ten per cent. in their wages," but conveniently forgetting that the same laborers had been stripped of 65 percent of their wage the previous year. A decidedly probusiness administration took power in 1878. Mayor R. R. Herrick, who inherited the growing debt of his predecessors, used his inaugural address to lambast the very notion of taxation. "Nothing so effectually strangles manufacturing and all business enterprises as a high rate of taxation," he argued, adding, "our resources are unlimited, and idleness cannot be long maintained." In this way, contemporaries grasped the connection between social problems and business, even if they shied away from making politically inconvenient associations.[24]

An anonymous writer to the *Leader* pointed out the discrepancies between public cost and private benefit. After reading the annual report of the city engineer, the writer noted that "three-fourths of the expense [of dredging the Cuyahoga] will be incurred above upper Central way bridge." "Everyone knows," the writer observed "that no one uses the river above the upper Central way bridge except [Standard Oil]." Even Mayor Herrick's own health department issued a report detailing how the "slow but continued contamination of our lake and streams of water from sewerage renders all additional deposits in them more objectionable." Despite such cries for the city government to force businesses to bear the burden of their own costs, Mayor Herrick refused to act. Instead, he used his position to advocate the creation of a private army, "subject to the call of the Mayor only," with its own fortified armory in downtown Cleveland where it could defend the capital against "cases of riot or insurrection" from the laboring classes. Cost was no barrier to the defense of private property.[25]

By 1880 the city government under the direction of Mayor Herrick served as little more than a tax-slashing organ of the Chamber of Commerce. Mayor Herrick warned against increasing public control, for "Capitalists in seeking a location in which to invest their money in business enterprises look as much to the rate of the tax levy at that place as to any other one thing, and will shun a city where the rate of taxation is excessive." "[O]ur aim," Herrick explained, was "to cut down and keep down the rate of taxation in this city as much as is consistent with its best interests." The health department of the city, according to its new chief, "was placed in the hands of a Board composed of Medical men . . . from education," who, instead of hunting down the sanitary abusers in industry, used their new station of legitimacy to initiate a witch hunt against midwives, whom they deemed "ignorant pretenders" of their own academy-earned education. In a meeting to address public outrage against the continued pollution of the city's water by industry, W. B. Rezner, the lone sanitary officer on the board, voiced his frustration. "If all the smell must be banished then the Standard Oil Company must move their works," Rezner argued. "I am in favor of regulating these concerns, and if the proprietors will not allow regulation, why I say abate their factories; but the board must give me power to do it, so that I will have something to fall back upon."[26]

The city government of Cleveland found itself in *the* dilemma of the Gilded Age. The very forces that built the city up from its modest frontier beginnings were poisoning its water and threatened to push society to the verge of class warfare. The solution each successive administration chose for this problem illustrates an abiding desire to accomplish two contradictory goals—protect the public welfare while not violating industry's sovereignty from government control. Water unified the landscape and revealed the contradictions of Gilded Age America. It connected the city's residents, rich and poor alike, to the environmental consequences of industrial wealth. The city government turned to ever more aggressive technological solutions for social problems.

Fearing that a raise in taxes would discourage industry, the city of Cleveland continued to borrow funds to build engineering systems that promised to deliver their citizens clean water. The chief civil engineer during Mayor Herrick's term, B. F. Morse, believed he found a solution to the city's woes. In his 1881 report to the city council, Morse outlined an elaborate technological landscape for the city's future health. Aside from an underground grid of interconnecting sewers, Morse entertained two possible systems for cleansing the Cuyahoga River. The first was a dam, "at some point above the city" that could "hold a large body of water," periodically released, so "the river can be flushed out occasionally." The second plan, which Morse believed more feasible, consisted of a "tunnel or conduit . . . constructed from the lake, under the city, to some point up the river," where the "lake water . . . discharging into the river at the upper end of the tunnel" would increase the river's natural flow and its ability to process wastes. "In this way the river could be flushed out," Morse concluded.[27]

Over the next decade, special committees met, drafted plans, and disbanded while the city continued to drink toxic water. Only a typhoid fever outbreak in 1890 spurred the city government to propose any meaningful change to the hydrology of the region. The outbreak struck in summer, low-water season for the river, and coincided with "a prolonged supply of muddy water." The health department suspected infectious waste had circulated through the city's waterworks. After all, they reasoned, "Other substances which are carried by river and sewers to the lake certainly come back to us in this way, and why not the virus of infectious diseases?" The city fathers' miracle tunnel, long supplementing the city diet with petroleum waste, now had an active hand in compromising the health of the very citizens it sought to defend. Despite the failure of hydraulic technology to deliver clean water, the city council replicated their previous mistake.[28]

By 1892 the council, with the backing of their health division, proposed a further extension of the lake tunnel. Meeting with a cacophony of questions and criticism, the health department used the occasion of its annual report to denounce the "opposition, even ridicule, from certain parties . . . to a measure so essential to the health of our people." Unwilling to contemplate any course other than the implementation of the technological tools at their disposal, the health officers declared that opposition to their plans "implies a degree of stupidity unworthy the present enlightened age." The conceit of their authority evidenced itself when they blamed the plan's detractors for their own failing, stating boldly "To oppose therefore the extension of the tunnel farther out in the lake, is virtually to advocate and encourage the exposure of the citizens of Cleveland to infectious diseases." Indeed, in 1896 the city's food inspector discovered the "tap water of the city contains from 300 to 2,000 families of bacteria per cubic centimeter" and found "without the question of a doubt, the presence of sewage matter" in all samples of the city's drinking water.[29]

By 1895 the city government commissioned a study by three expert engineers to consider the manifold projects of an intercepting sewer system, extension of the lake tunnel, and a river flushing tunnel. The expert commission delivered three conclusions in their special report to the mayor and city council. First, any future water intake in the lake "should not be less than ten miles" from the mouth of the Cuyahoga to avoid contamination. Second, the combined waste dumped into the river from citizens and industry had combined to create a subsurface "sludge, which in warm weather becomes a powerful agent in still further polluting the river by putrefactive fermentation," and "becomes perceptible by the gas bubbles on the river surface . . . and by an offensive odor when the water in time of drought has become black in color and saturated with filth." Last, the expert panel recommended that the "only course left . . . is to supply the deficient flow [of the river] by pumping [water] through a tunnel from the lake to the nearest point on the river." Fortunately for the taxpayers of Cleveland, the flushing tunnel— estimated in 1897 to cost one million dollars—never made the transition from pipe dream to reality.[30]

John D. Rockefeller, known primarily to American historians for his zeal to minimize costs through vertical integration, could have avoided the entire episode by displaying a modicum of his famous parsimonious character. The acid-laden water his refineries loosed into the lake via the Cuyahoga could have been recovered. In response to charges leveled against the acid manufacturers of the city, owners of the Acid Restoring Works wrote to the *Leader* as early as January 1868, describing their own process of recovering used acid. "The acid used in discoloring petroleum oil is afterwards run into lead-lined tanks," they wrote, "instead of being run into the river as formerly; from these tanks we convey it in a tank wagon to the works." They describe a simple system of separating the petroleum remnants from the acid, which is finally recovered in pure form after allowing water to evaporate in the open air. Toward the close of the nineteenth century Standard Oil would eventually begin such recycling efforts, for its own internal ledgers revealed that, on a year-to-year basis, acid cost more than labor. In Cleveland the river, the lake, and ultimately the citizens proved a far cheaper method of disposing of waste than constructing the proper recovery equipment. In the long term, the savings accrued through recycling acid would have partially offset the initial investment in a central recovery station or in site-specific works located at each refinery.[31]

TUNNEL VISION

The increased water supply created a new culture of water consumption in the closing years of the nineteenth century. When the first waterworks opened in 1856, the city boasted a population nearly equivalent with its daily consumption in gallons—thirty-eight thousand—or one gallon per person, per day. Many citizens still relied on their own wells or barreled lake water for their needs. Some, including a teenage Rockefeller, still bathed in the Cuyahoga and Lake Erie. However, the introduction of the bountiful lake water, made instantly available with the flush of a toilet or the twist of a tap, introduced what one water historian has called "dis-economies" of water consumption. During the summer of 1885 consumption exceeded supply, forcing several factories to close temporarily. The Cleveland Board of Councilmen, noting the prodigal use of water for such extravagances as "the sprinkling of streets and lawns," urged the city commissioners to enforce a water ordinance. Water, like fire, had been transformed from a vital natural force into a mere resource. The ties between a pint of water and the environment began to fade for Americans once technology relieved them of the labor necessary to fetch a pail from a well, river, or lake.[32]

With every new waterworks extension in the search for clean water, the city's engineers unknowingly calibrated the city's population to ever-higher levels of consumption. The waterworks that had distributed 38,000 gallons a day in 1856 delivered 3,746,907 gallons in 1871 or nearly 40 gallons per person, per day. Twenty years later, although the city's population had nearly tripled—from 92,829 in 1870 to 261,353 in 1890—daily water usage exploded by a factor of 8.7, rising from a daily consumption of 3.7 million gallons in 1871 to 32.2 million gallons in 1891. A year later, the city's waterworks commissioners noted that the increase in water use for 1892 over the previous year proved more than the "entire consumption for the year 1871." The search for a technological fix to the problem of water pollution in Cleveland set the city down a course of supplying prodigious quantities of water to citizens and businesses willing to find ever more uses for it while doing little to address the initial cause for concern—quality.[33]

The city council offered Cleveland the final, costly solution to water pollution at the turn of the twentieth century. Despite the expert panel's previous warning against building a lake intake within ten miles of the Cuyahoga's mouth, the council authorized construction of a massive, nine-foot-diameter tunnel in 1896. Completed in 1904, the tunnel extended four miles into the lake. By 1906 the debt for the continued expansion of the waterworks system stood at $4,266,000 and would grow sixfold over the next quarter century.[34]

Standard Oil, while not alone in contributing water pollution to the Cuyahoga River and Lake Erie, had necessitated the expansion of the waterworks system at the close of the Civil War. Rockefeller himself failed to contribute a dime in taxes to its maintenance. When Cuyahoga County threatened a suit against the $310,000,000 in property he possessed at his estate on "Millionaires' Row," he fled the city limits for an East Cleveland suburb and built a second, larger estate at Forest Hill. These events would still stir him years later. In his famous interviews with William O. Inglis during the late 1910s, Rockefeller referred to tax collectors as "blackmailers." Rockefeller's legacy of philanthropy, in fact, should reinforce this image of a Gilded Age culture that held individuals responsible for their own success and misfortune. Rockefeller and those sharing his ideals believed that public welfare was a voluntary burden for those in society with the ability and virtue to support it. It appears, however, Rockefeller had no objection to the taxpayers of Cleveland bearing the costs for his refusal to recycle his own acids for more than a generation.[35]

Simultaneous with the extension of public infrastructure to bear the private costs of industry was the further retrenchment of capital in the Forest City. Following another prolonged economic depression in the mid-1890s, business again cut labor costs, which resulted in an organized boycott of local industry by Cleveland's labor movement. The chamber of commerce, contemplating the use of their private army against organized labor, served as a meeting place of capitalists in late summer of 1899. "The object of the

The new crib is towed out to the lake in 1898. One of the temporary cribs used before the tunnel was completed can be seen in the background to the right.

Credit: Cleveland, Ohio, *Annual Report, 1898*, xiv.

meeting," according to the observing *Leader* reporter on August 9, 1899, "is to take steps to crush the boycott." Collecting 312 endorsements from local business owners, the meeting complained that the boycott was "restraining citizens and business men by intimidation from exercising their individual liberty," and concluded that "such a condition is un-American, as well as detrimental to the good name of our city." A day later, on August 10, the meeting drew 500 "citizens," who, "Almost without exception . . . were men of prominence in the business community." This second meeting elected J. G. W. Cowles—Rockefeller's personal real estate agent—as chairman, now given responsibility to collect subscriptions to break the strike.[36]

No subscriptions were raised, however, against the city government or commercial interests of the city when a boiler fire ignited a pocket of gas and destroyed the temporary wooden crib that housed the lake tunnel workers in the summer of 1901. The force of the blast threw men and steel cylinders forty feet into the air. As lake waters and mud rushed into the excavation tunnel, men who survived the explosion struggled to tread water five miles from shore. The surviving workers spent the night trying to signal the shore until the next morning when a tug arrived to carry them to safety. James Williams, one of the five men killed in the accident, had only a week earlier survived another gas explosion that claimed nearly a dozen lives under the lake. Williams had descended into the waterworks tunnel to save men over-

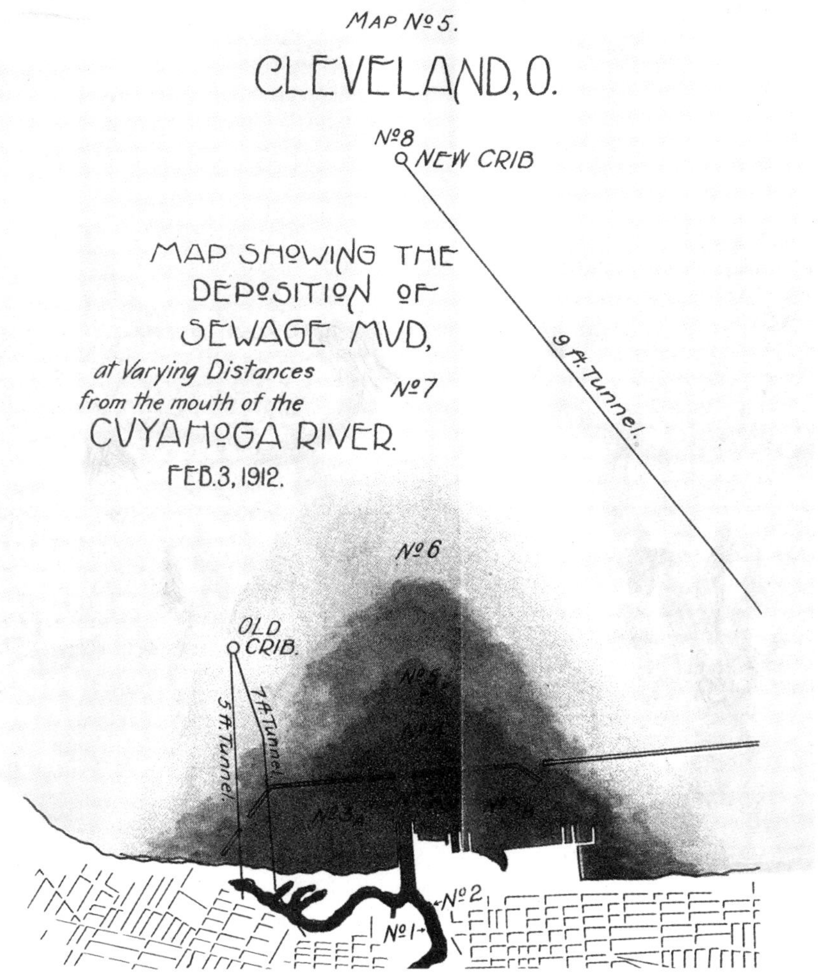

MAP Nº 5.

CLEVELAND, O.

Nº8
NEW CRIB

MAP SHOWING THE
DEPOSITION OF
SEWAGE MUD,
at Varying Distances
from the mouth of the   Nº7
CUYAHOGA RIVER.
FEB.3, 1912.

Nº 6

OLD
CRIB.

9 Ft. Tunnel.

5 Ft. Tunnel.

7 Ft. Tunnel.

Nº 2

Nº 1

A study from February 1912 shows the relationship between the city's water pollution and waterworks infrastructure.

Credit: Jackson, *Sanitary Condition of the Cleveland Water Supply.*

come by gas fumes. By the time it opened in 1904, thirty-seven workers had died during the construction of the tunnel.[37]

Even after the city shifted to a Progressive mind-set, with the election of Mayor Tom Johnson in the first year of the twentieth century, it remained committed to a water infrastructure predicated on the logic of the Gilded Age. In a report on the water quality of the city's new crib in 1905, the sanitary engineer George C. Whipple noted that although rates of typhoid dropped with the transition to the new crib in 1904, traces of pollution still found its way into the city's water supply. The new lake tunnel might only offer a temporary solution to the city's water woes. By 1913 rates of infectious disease had increased (typhoid rates jumped from 5.9 per 100,000 residents in 1912 to 13.5 per 100,000 a year later) and the secretary of the filtration committee, R. Winthrop Pratt, reported, "the public water-supply of Cleveland has been continuously very objectionable."[38]

The city's response was to treat the water with hypochlorite of lime, set aside $1,000,000 for an expanded intercepting sewer to dump wastewater directly into the lake rather than the Cuyahoga, and plan for the construction of a sewage treatment plant and water filtration works. These decisions set the city on a course of managing the region's hydrology that became a hallmark of the modern city in the twentieth century. The technological infrastructure became a necessity as the population topped a half million by 1910. Historians Thomas P. Hughes and Christopher Jones have argued that technologies requiring heavy initial investment and continuing maintenance costs soon acquire their own momentum. Earlier commitments of time and wealth compound, pricing out alternatives, and a parallel social system, or bureaucracy, gains political power to determine the infrastructure of entire cities. As technology replaced rivers, Cleveland traded one form of dependence with another, raising the social and environmental stakes along the way. The city's waterworks stood as a grim monument to the early decision to clean up private costs with public infrastructure in the Gilded Age.[39]

CONCLUSION

Northeastern Ohio's hydrological order, relatively unchanged since the last glaciation ten thousand years ago, became a rationalized system of canals, pumps, and subterranean sewers. With a mere thirteen miles of pipes distributing just over 127 million gallons a year in 1857, the quest for clean water led the city to build an additional 112 miles of pipe distributing over 10 million gallons a day twenty-three years later. Per capita consumption rocketed up from 7.75 gallons to 65.25 gallons a day during this time of declining water quality and booming population. The failure to arrest the problem of pollution at the source delivered the city into an industrial analogue to the punishment of Tantalus. Hydraulic technology intended to treat the symptoms rather than the cause of problems would ensure poor drinking water, which in turn would justify the need of further technology.[40]

These Gilded Age developments contributed to the creation of water crises that would plague the industrial metropolis in the twentieth century. Interlinked with its surrounding ecosystem through a technological system comprehensible to only a small class of specialists, Cleveland entered the new century unable to supply its burgeoning population with the clean water that lapped at its coastline. As urban historian Martin Melosi has argued, technological reforms created both short- and long-term effects. In the short term, technological systems solved immediate problems, whether political or environmental in nature. "In the long run," according to Melosi, "the emphasis on project design rather than careful planning often focused attention on the most immediate goals of implementation rather than the potential resilience of the system or its capacity to adapt to the pressures of growth." Despite the construction of a centralized water system maintained by sanitary engineers, technology failed to deliver the city from the environmental costs of industrialism. For Clevelanders at the end of the kerosene age, the lament of the Ancient Mariner reflected their own experience:

> Water, water, every where,
> Nor any drop to drink.[41]

FOUR

W. H. Foster, manager of Standard Oil's No. 2 works, found himself in a peculiar position in 1898. Foster was arrested for doing his job—operating a refinery. Of the 192 notices served in 1898 to various industries by James McLaren, Cleveland's public smoke inspector, only those issued to Standard Oil precipitated further visits and, as Mr. Foster knew all too well, a single arrest. While Cleveland's city government could engineer a new water system for the city, cleaning the atmosphere lay beyond the reach of technology. On the back of a growing civic concern over the quality of air in the industrial metropolis, Cleveland's city council created a modest regulatory body with an aim to curb excess pollution and sate public anger.[1]

Throughout the Gilded Age, the city's air remained dirty, at times even dangerous to breathe. While the city experimented with regulation, the rich fled to the clear skies and clean streets of new suburbs. The city's working communities, forced to live with the environmental consequences of industrialism, accused the municipal government of supporting anti-American notions of social privilege, even class warfare. In Cleveland the atmosphere, like the public waters, served as an early battleground for the social struggles that historians associate with the beginning of the reforms of the Progressive Era.

This struggle extended to land as Clevelanders identified parkland as sanctuaries from urban pollution. A political battle over control of Cleveland's growing park system developed in the final years of the nineteenth century. As the rich relocated to the suburbs, they donated thousands of acres of parkland to the city. Conflict emerged when Cleveland's citizens noticed how these distant parks confiscated significant municipal taxes and increased land values for surrounding suburban real estate developments. In its relationship to the growing park movement, air pollution's geographic dimension precipitated social conflict that propelled Cleveland into the Progressive Era. Air currents dictated a new landscape of social conflict, where individual wealth—not public goals—determined health. Standard Oil's refineries were among the greatest contributors to Cleveland's air troubles and John D. Rockefeller played a leading role in forging a new geography of health in the city's new suburbs.[2]

THE ALTAR OF INDUSTRY

The coming of air pollution to the American urban landscape was at first welcomed as a sign of material progress. At the close of the Civil War, when so much of the nation's wealth had been mobilized or destroyed in a half decade of carnage, industry served as a positive focus for a war-scarred nation. In the final year of the war, Cleveland mayor George Senter mentioned with pride the "gratifying fact that the manufacturing interests and the carrying trade . . . are rapidly enlarging themselves." Industry offered a rebirth from the largest setback the young republic had yet faced. Thus, it is not surprising that Chicago industrialist W. P. Rend wrote in 1892 of the smoke that belched from the Windy City's factories as "the incense burning on the altars of industry." After a century of struggle with an untamed environment, Rend saw in Chicago's pollution a signal "that men are changing the merely potential forces of nature into articles of comfort for humanity."[3]

Following the Civil War, many took comfort in a resumption of America's Manifest Destiny. Cleveland businesses refocused their war spirit by dredging the Cuyahoga, laying railroad trunk lines, and building a viaduct over the river. Already a center of commerce because of the canal connecting it to the Ohio River and the Great Lakes, Cleveland saw its industry flourish in the years following the Civil War as new enterprises in paint (Sherwin Williams, 1866), iron ore (M. A. Hanna Company and Globe Iron Works, mid-1860s), and petroleum refining (Standard Oil Company, 1870) sprang up along the banks of the Cuyahoga. These new industries, however, would also transform the city's atmosphere.[4]

However sweet the smell of industry was to the nostrils of Cleveland's industrialists, it was certainly not good for them. The primary fuel source for Cleveland's growing industry following the Civil War, bituminous coal, released a potpourri of toxic vapors when burned in the boilers below the oil stills, in smelting furnaces, or in domestic stoves. Cleveland citizens did not require an official report to inform them that the air they breathed was harmful. Although the vapors' chemical properties were not fully understood until the middle of the twentieth century, the new industrial pollution created observable effects on plant and animal life.[5]

The two most severe by-products of burning fossil fuels—sulfur dioxide (SO_2) and carbon monoxide (CO)—work against biological systems in malicious ways. Sulfur dioxide, when released into the atmosphere, oxidizes and binds to water molecules to create sulfuric acid. In high concentrations, the acid is absorbed through the stomata of leaves and causes the plant cells at first to deactivate, turn yellow then brown, and eventually burst—robbing the plant of the ability to photosynthesize. Sulfur dioxide, when inhaled by animals, causes respiratory irritation, increased mucus production, and, when present in high concentration in the atmosphere causes any watery membranes—such as the eyes—to become inflamed. With the combustion of large amounts of fossil fuels, sulfur concentrations become so high as to cause a regional change in atmospheric acidity commonly referred to as "acid rain," which not only destroys plant life but even melts away the façades of stone buildings and statues.[6]

Carbon monoxide works against animal health by attaching to the oxygen-bearing protein hemoglobin in blood, robbing tissues of oxygen in proportion to the level of carbon monoxide in the atmosphere. In high enough concentrations, the body lacks enough oxygen to perform simple brain functions or operate organs, leading to loss of consciousness, brain damage, and even death. The atmosphere in communities downwind from refineries, under the right conditions, could bottle up fossil fuel emissions and create a host of respiratory problems.[7]

Cleveland's industries—especially the coal-consuming iron mills and oil refineries—emitted these pollutants in prodigious amounts during the industrial boom that followed the Civil War. In September 1872, *Harper's New Monthly* featured a travel narrative of the novelist Constance F. Woolson that captured these changes. Beginning in Buffalo, Woolson toured the Great Lakes in the vessel *Columbia* and reached Cleveland's shores soon after. She painted her impression of the Forest City in striking tones. "Presently the spires of Cleveland came into sight, a cloud of smoke resting over the city coming from iron-mills and oil-refineries crowded together on the marshy flat of the Cuyahoga Valley." The ship's well-heeled passengers crowded the deck of the *Columbia* to catch a glimpse of the city and inquired of the skipper "What kind of place is Cleveland, captain?" "Good enough place; 'ily, though." The curious passengers further questioned the captain, whose Irish accent Woolson retained:

> "Captain, what are the contents of those barrels?" inquired the Utica schoolmistress who presided over the band of school girls.
> "Ile, marm."
> "Captain, pray what is this disagreeable odor?" said Mrs. Peyton, taking out a vinaigrette.
> "Ile, marm."
> "What makes the water look so funny?" said Curleylocks, one of the schoolgirls, gazing over the side.
> "Ile, miss."[8]

Given until sunset, the passengers disembarked to visit the city. Many wished to tour Euclid Avenue, "where the big houses are," but one passenger in Woolson's carriage demanded they first visit the refineries. Despite pleas to not "go near that oil; it will give us all headaches," Morris—a stubborn passenger—insisted, stating they "could see residences anywhere, but Cleveland is a great oil place: you may call it 'highly refined.'" The touring carriage convinced a passenger, Major Archer, who "had a friend in the business," and the tour began. "The flat was crowded, odoriferous, and smoky, with lumber, oil, and iron; but the oil predominated. Blue barrels met our eyes on every side, huge tanks rose from the ground like fortifications, and a network of pipes, elevated high in the air, ran hither and thither, while over, under, and throughout all the pungent petroleum made itself felt in every breath we drew. On we went, and the smoking chimneys grew into a forest." Major Archer's friend was undoubtedly a member of the Standard Oil Company for blue barrels were the calling card of the company until pipelines and bulk transport made them obsolete later in the century. The group concluded their tour of the refinery—one of the few descriptions of the interior of Standard's Cleveland Works of the time—and made their way toward Euclid Avenue, which the industrialists of the city called home. The party, astonished by the contrast with the bleak Cuyahoga, marveled at the "velvet lawns, conservatories, shrubbery, statues, and fountains of these fine residences." Even the "noble trees" of the Forest City earned their praise as they watched the sunset on the lakeshore and repaired to their ship to continue the voyage.[9] Unlike the affluent passengers of the *Columbia*, the people of Cleveland could not abandon the toxic atmosphere created in the industrial flats. In the half century after the advent of the kerosene age, the city of Cleveland awakened to a new ecology. Like so much of Gilded Age America, the new urban environment favored the rich over the poor.

MAKE A STINK

As Clevelanders opened their windows during the hot summer of 1870, more than the night breeze entered their homes. "The rancid, suffocating breath of the oil tanks again makes the hot nights a burthen to us," the editor of the *Leader* announced in late July 1870. "Soon after dusk the foul vapor comes pouring down the valley of the river, and . . . it is so dense in the lower part of the city that persons whose lungs are at all delicate find it next to impossible to sleep." Although journalists at the *Leader* came to industry's defense when labor unions and progressive city governments sought to exert more democratic control over business, the stink that wafted into their bedrooms at night led writers to call on the city council to immediately "come to the rescue of the community."[10]

After early reports suggested the chemical manufacturers were the source of the stench, *Leader* reporters discovered that "the Acid Restoring Company [had] been removed eight miles away" following a previous outcry. In its absence, however, "certain oil refiners within the city limits have undertaken the process which creates the stink." The *Leader* dispatched a reporter to hunt down the source on an overnight trek. During his journey, the reporter "could hear the slamming down of windows by disgusted residents of Euclid avenue, as the stench penetrated within and aroused the sickened occupants from sleep." His overnight adventure would include several acts of trespassing—a bold case of investigative journalism and a testament to the severity of the problem. His journey ended at the Walworth Run where "the rank oil smells and the sharp flavor from the acid works" mingled with the noxious "perfume" of the tributary.[11]

The *Leader* vacillated between suggesting that abatement would cause Cleveland to get "laughed at by all her sister cities" and calls for fresh air no matter the cost. "It is probably desirable to some of our citizens to have the acids refined," the *Leader* stated after their nocturnal investigation, "but it is not the just privilege to suffocate their fellow beings in accomplishing that purpose." Having sought salvation in technology for the city's water woes, could the government engineer their way out of this problem?[12]

Cleveland's government had acted as early as February 1860, less than a year after the Drake Well strike, to ban the refining of coal oil within the city limits, an action the *Leader* opposed at the time. By 1870, however, the *Leader* was echoing a much louder public sentiment against the oil refineries. Forced to locate outside the city's limits, the refineries sprang up along the shores of two tributaries of the Cuyahoga—the Kingsbury and Walworth Runs. Located on opposite sides of the river, these streams soon became known for their stench. The *Leader* dipped into literary imagery, calling the Walworth Run "The Slough of Despond." The people forced to live in the neighborhoods around these industrial sinks responded through civic action. Upon hearing of plans for a new oil refinery, laborers of the city's Seventh Ward circulated a petition against its construction and passed the petition on to the city council in the summer of 1875. Dr. Thayer, of the board of health, lent his support to their cause, "believing that [the refinery's] fumes would endanger the health of the people."[13]

Despite these localized events, no comprehensive plan guided the city government officials, who preferred to address problems as they materialized on a case-by-case basis. By the late nineteenth century, the officials' inaction became apparent to everyone as the size and scale of refineries grew in the city. The complaints of a few impoverished wards spread to the entire metropolis as pollutants refused to respect property lines, and concerned citizens challenged the fundamental defense made by business. "They talk about driving capital out of the city if these nuisances are forced to abate," T. H. Lamson wrote in to the *Leader* in 1880. "More capital is kept out than is invested in all of them; and further, the desirability of Cleveland as a place of residence is seriously endangered." Lamson's words proved prophetic as the affluent abandoned Millionaires' Row for distant suburbs.[14]

THE FOREST CITY

Only when the trees that gave Cleveland its nickname began to die did the city council finally take notice. No amount of pressure from their constituents or the trained medical doctors at their own board of health spurred as much action as a few dead shade trees. Working-class neighborhoods had been denuded earlier in the century—if not by the violence of pollution, most certainly by cold hands searching for fuel. Only Public Square, the few parks, cemeteries, and of course the verdant pastoral estates of Millionaires' Row contained ample trees. What is certain is that by 1891 the death of the Forest City's trees moved the city council to elicit a study and recommendation from the US Department of Agriculture.

As chief of the Forestry Division of the USDA, B. E. Fernow, a founder of American forestry science, and J. C. Arthur, professor of botany at Purdue University, penned two reports to Cleveland's city council. In his investigation of the "rapid decline and death" of the "shade trees of the city," Arthur found that "such cause is to be found in the smoke from the large manufacturing establishments, and especially from the oil refineries." Although he lacked the equipment to measure air quality, Arthur confirmed, "from the well-known abundance of sulphur in American soft coal and crude petroleum, there can be no reasonable doubt that it occurs in sufficient amount to largely or wholly account for the destruction of trees."[15] After his visit to Cleveland, Arthur "found the trees of the city in a really deplorable condition," and the "largest trees in the central part of the city . . . without exception, greatly enfeebled and slowly but surely dying." As workers removed dead or dying trees from parks and Public Square, Arthur had the opportunity "to examine the roots, interior of the trunks and the highest branches of some of the largest trees in the last stages of the injury." "I am fully convinced," Arthur concluded, "that the chief difficulty with these trees arises from the befouled condition of the atmosphere."[16]

Fernow's visit to the city gave him confidence in stating, "Death occurs from acid poisoning." After an analysis of samples, Fernow concluded "the sulphurous acid [was] desiccating and destroying the tissue of the leaf." Arthur confirmed the suspicions of Cleveland residents when he discussed the source of toxins: "The great amount of soft coal used for heating and in the production of power by the large manufacturing establishments, and of crude petroleum burned at the refineries, gives rise to enormous quantities of smoke which darkens the atmosphere and covers trees, buildings and other objects with a thick coating of soot. In addition to the soot, the burning of coal and crude petroleum gives rise to a number of gases, such as carbon monoxide, sulphurous acid and arsenious acid." Fernow offered the city council two possible paths to take: "either to prevent the escape of the noxious gases or else to plant only such trees as are exempt, or partially so, from ill effects of the gases." However, he made clear to the council that only one true choice lay before them, for in suggesting hardy tree species he sounded a pessimistic tone, stating "it may be said that where the smoke nuisance is excessive none will survive." Even if "effective measures were taken at once to suppress the smoke nuisance," it would only be "possible to save, by judicious pruning, such trees as are not too badly affected, and have retained vigor for reproduction." As Fernow concluded his report, he wrote, "The City of Cleveland, beautiful and attractive through its verdure of luxuriant trees, may well be alarmed at her loss and stand ready to guard her trees with jealous care." The response from the council would serve as a rare experiment in public oversight of industry in the Gilded Age.[17]

Despite testing the legal boundaries of business in the Gilded Age, no

This photograph taken in the early 1900s shows a working-class community over-looking the Cleveland Flats near the heart of the oil district. Miles from the luxurious suburban parks, this image illustrates the class boundaries of the city's geography of health. The white laundry contrasts with the accumulated soot on surrounding buildings.

Credit: The Western Reserve Historical Society, Cleveland, Ohio. Cleveland Picture File 1 (file 8686).

member of the Standard Oil Company served a single day behind bars until inspector James McLaren collared W. H. Foster in 1898. McLaren's action did not occur without precedent. The city council passed a smoke ordinance as early as 1882, and John Vandevelde—the first smoke inspector—policed Cleveland's sky a year later. In his first year, Vandevelde kept busy, making 1,163 visits to follow up on nuisance complaints and serving 535 notices to polluting factories. His job was an unenviable one. "So great was the prejudice against it," Vandevelde wrote in his first annual report, "and so strong the belief that it was impracticable, that an effort was made to repeal the ordinance." It is easy to sense how unpopular the new smoke inspector was in Mayor John H. Farley's annual message to the city council in 1884. "[W]hile much care should be taken to keep the city as clean and healthy as possible," Mayor Farley wrote, "care should also be taken to protect the people from the wiles and dilusions [*sic*] of hygienic cranks." Unlike Vandevelde or the sanitarians in the city's department of health, Mayor Farley had to answer to the businesses and affluent citizens who protested any increase in taxation.[18]

Vandevelde and subsequent smoke inspectors urged polluters to experiment with a bevy of filters, flues, sprinkler systems, and chimney designs in an effort to clean the city's air. In his first year of work alone, Vandevelde was able to convince Standard Oil, "with their hundreds of furnaces," to try "most all the appliances in the market." Despite a decade of experimentation, ineffective devices or a lack of diligence in their application left Cleveland's skies as dirty as ever, as the dying trees demonstrated. By 1893 the new smoke inspector, George F. Leick, wrote in his annual report to the city council that the "market is flooded with numberless devices for consuming and preventing smoke; the great majority of these contrivances are absolutely worthless." Even more troubling, Leick reported to the council that the toxic air of Cleveland was taking a toll on more than just the city's trees. People were dropping in the streets where carbon monoxide levels concentrated to dangerous levels. "During the past year," Leick wrote, "this department paid out $1,057 to various hospitals for treatment of emergency cases picked up by the policemen on the streets." Doubting any abatement in the smoke nuisance, Leick believed the only responsible course lay in providing public relief by "the addition or setting aside of a ward in the City Hospital where this class of cases can be treated."[19]

Such a hospital already existed just downwind from the Flats industrial region. St. Alexis Hospital, founded in 1884, was located only a few thousand feet to the southeast of the main Standard Oil works and served victims of industrial accidents and respiratory cases. By the mid-1920s, when students of the Case School of Applied Science conducted the first study of Cleveland's air quality, the measurements taken from the roof of the hospital revealed the highest levels of atmospheric pollution in the city. The hospital averaged a stunning 2,038 tons of soot per square mile per year—four to eight times higher than the average of the British industrial cities of Leeds and Manchester (240–539 tons/sq. mi./yr.)—and even topped the worst areas of America's air pollution capital, Pittsburgh (1,950 tons/sq. mi./yr.). At that rate, the upwind smokestacks belched out and dumped over an acre-foot of soot on the square mile around the hospital every year.[20]

CASTLES IN THE AIR

By the 1890s the waters of Kingsbury and Walworth Runs had been saturated with waste to such an extent that they also contributed to the air nuisance, as the slurry of slop fermented and released vapors into the atmosphere. The board of health conveyed the urgency of the problem to the public. "All along [Walworth Run] the banks are necessarily soaked with [industrial waste], undergoing decomposition, and must be unhealthy."[21] As early as 1873, a special committee "prepare[d] a careful report, embracing the causes which operate to render Walworth Run an obnoxious and filthy nuisance" and concluded that these causes were the "refuse from seven oil refineries,

Culverting the Walworth Run, 1897. Evidence of the city's dedication to fix environmental problems with engineering technology rather than regulate business.

Credit: Cleveland, Ohio, *Annual Report, 1897*, xxx.

offal from eight slaughtering establishments, two soap factories, one stock yard, two city sewers, one distillery, two woolen mills, four breweries and two tanneries, any one of which is, in our opinion, sufficient to contaminate and pollute the water." Although the committee specifically noted that such dumping "is strictly forbidden by the State law of Ohio," its members could not but view the problem with the guiding ethos of laissez-faire culture. "But to prevent the various manufacturing and other establishments from discharging their refuse water into the Run," the special committee concluded, "would be a great injustice, as it would virtually cause them to close their business, and throw out of employment a large body of mechanics." Thus, even the guardians of public health concluded that pollution had become an existential right for business in the Gilded Age.[22]

The committee members' solution was to refine the ecology of the Cuyahoga watershed at taxpayer expense. "We would therefore recommend," the committee concluded, "that the stream . . . be allowed to remain (for the present) 'in statue quo,' and that a closed sewer be constructed so as to occupy a place in the valley and receive the refuse matter from those establishments and sewers now draining into Walworth Run." Given their reaction to the fouled waters of the Cuyahoga and Lake Erie, the council members' response was not surprising: both the Kingsbury and Walworth Runs would be culverted—at great public expense (the Walworth sewer alone cost an estimated $730,000)—and transformed into subterranean sewers to trap the stench on its journey to the river.[23]

In 1875, long before construction began on the Walworth Run sewer, the board of health reported to the city council that much of the organic "offal" dumped into the Walworth and Kingsbury Runs "is usually carted away by farmers or the fertilizing companies" to enrich the soil of the garden farms that ringed the metropolis. By 1878 the board of health urged that the "proper way would be to send teams to the various provision houses to carry all waste animal and vegetable matter direct to the country," but the council members balked at dedicating funds to "such an arrangement," because it "would involve increased expense, and that the city is evidently not in favor of." Because of the city government's unwillingness to subsidize such recycling efforts, the board of health "recommended that every household burn as much as possible of its kitchen garbage when other means of disposal are not at hand," further burdening the city's atmosphere. The death of the feedback loop between local farms and the city's waste represents one of the many unheralded casualties of the kerosene age alienating urban dwellers from the land.[24]

It should come as little surprise that those with an interest in low production costs abhorred the threat of municipal oversight. As citizens protested the continued contamination of their atmosphere, Cleveland's industrial leaders found ways to confront the rising tide of public interest in air quality. Finding the prohibitive ordinances lacking in severe penalties, most of them simply ignored them—applying the logic that if the costs of noncompliance with the law proved lower than cleaning up their operation, they had an obligation to their shareholders to take the path of least resistance. In 1880 health officer W. B. Rezner expressed frustration at the lawless nature of Cleveland industry during a public meeting concerning the stench from a local slaughterhouse: "[H]ow under the sun am I to abate them. I served a notice on them this morning, and have given them twenty-four hours to cease manufacturing smells. If I find them still creating a nuisance tomorrow I shall swear out a warrant and have them arrested. Judge Young can impose a fine, and they can go on their way rejoicing. I am powerless to proceed differently." When the meeting turned to oil refineries, Rezner revealed how he lacked the power to confront the source of pollutions. He understood the concerns of industry, stating that if "all the smell must be banished then the Standard Oil Company must move their works and Walworth run from end to end be devastated of manufactories." However, a complete elimination of pollutants was never the goal of the smoke ordinances. "I am in favor of regulating these concerns," Rezner stated, "and if the proprietors will not allow regulation, why I say abate their factories." Rezner pointed to the true source of inaction by declaring, "the board must give me power to do it, so that I will have something to fall back upon." Just as the modern environmental movement has learned, Rezner recognized that prohibitive laws meant little if no one could enforce them.[25]

As was the case with their reaction to water pollution, the city council members chose solutions to the air nuisance that gave them the appearance of acting in the public's interest (authorizing ordinances and establishing the office of smoke inspector), but they failed to invest their actions with sufficient force to produce meaningful results. With the smoke inspector on the beat, the council could dismiss any criticism as the ravings of impatient radicals or assign blame to the smoke inspector himself. Some residents, like the anonymous "Sufferer," wrote to newspapers calling for an end to the posturing by business and city government alike. During the summer of 1880, the anonymous writer complained, "night after night, when the air was close and stifling, have we been awakened from a sound sleep by inhaling the noxious odors from these rendering establishments." The writer had "been obliged to close every window to keep, if possible, some of it out," because "the odor is so strong that it often brings on cases of severe vomiting." Harking back to the days of Benjamin Franklin, the writer believed that Cleveland "from its natural advantages should be, the healthiest [city] in the United States." This "Sufferer" concluded by asking how the city's public officials could permit such violations. "Do our health officers understand that the people are looking at them anxiously to have it abated? They are paid salaries for the performance of their duties, and if they have not got the nerve and backbone to fulfil [sic] them, let them step out and others go in that will." The city had dissembled, but Clevelanders had seen through the smokescreen. Without an engineering solution to the problem of air pollution on the horizon, rich residents would begin to vote with their feet.[26]

SUBURBAN AIR

"Pure air is good food" declared health officer G. C. Ashmun in his report to the city council in 1881. After spending many summer nights trapped inside their spacious homes on Euclid Avenue, the affluent residents of Millionaires' Row began an exodus to the suburbs rather than breathe air that caused them to become ill. T. H. Lamson's warning in 1880 that "the desirability of Cleveland as a place of residence is seriously endangered" came to fruition. Samuel Andrews, the English chemical engineer who managed Standard's stills through their formative years, built a thirty-three-room castle on Euclid Avenue, stocked with furniture and carpets custom-made in England and maintained by an army of a hundred servants. However, Andrews and his family abandoned it after living there for only thirteen years, shuttering the Victorian mansion in 1898. As an original member of the Standard Oil Company, Andrews could throw caution to the wind and relocate when it grew foul.[27]

Rockefeller himself had abandoned his relatively modest residence on Euclid Avenue where he lived close to his business partner Henry Flagler. He moved six miles east to a several-hundred-acre Forest Hill estate, far removed from the bustle and stench of the city. On a hill commanding a view of the city and Lake Erie, Rockefeller built a stately Victorian mansion that was originally planned as a health resort, and which he named "The Homestead." According to one of his own employees, Rockefeller harbored an obsession with clean air. "He was also very particular to see that the offices were well ventilated as he believed in plenty of fresh air," the anonymous employee recalled later in life. "[W]henever he stepped into an office where the air was not as fresh as he thought it should be he did not hesitate to make the fact known and arrange for such ventilation as he considered necessary."[28]

When not at work, Rockefeller committed his energy to controlling every aspect of his new estate. He planned pathways and flower gardens and even selected individual trees at his new suburban home. "Of all the profitable things which develop quickly under the hand," Rockefeller wrote after retiring from the oil industry he had built, "I have thought my young nurseries show the greatest yield."[29] Like many conservationists at the turn of the century, Rockefeller viewed open spaces as essential to the health of city dwellers, even directing the then thirteen-year-old John Jr. to take advantage of the fresh atmosphere of Forest Hill during an illness in 1887, thus revealing his understanding of the connection between health and environment. "Be sure to take good care of your health," the elder Rockefeller directed his son in 1899. "This is of the first consideration." Rockefeller also understood the commercial value of fresh air. In the summer of 1888 his brother Frank Rockefeller informed him of the potential for adding more land to his real estate holdings in suburban East Cleveland. Urban pollution created an opportunity for investment, Frank wrote, because "the lots may be higher [in price] owing to the electric road building, and the number of first-class families seeking fresh air in that direction."[30]

The consequences of refining petroleum inscribed a new geography on cities like Cleveland. The urban core retained heavy industry saddled by ethnic enclaves with names like Whiskey Island, Little Cuba, and Dopetown, while affluent residents retreated to suburbs at the edge of the city.

Samuel Andrews's Euclid Avenue "Castle" known as "Andrews' Folly" after the palatial dwelling was abandoned in favor of East Cleveland suburbs far removed from industrial pollution.

Credit: The Western Reserve Historical Society, Cleveland, Ohio. Cleveland Picture File 1 (file 6317).

Early critics suggested that Rockefeller's legacy of philanthropy amounted to an apology for the destruction wrought during his business career. In some cases, philanthropy reinforced inequalities. The parklands created by suburban residents such as Rockefeller served as a barrier between the affluent and the urban blight they fled. The efforts of a handful of wealthy residents of Cleveland's new East Side suburbs led to municipal ownership of parkland stretching from the lakeshore to Shaker Heights—a six-mile strip of nearly uninterrupted parkland. A hundred years to the day after the founding of the city by Moses Cleaveland on July 22, 1796, Rockefeller gifted 276 acres of land along Doan Brook, which completed a strip of parklands from the lakeshore up to the Shaker Heights suburb. In a letter to the city council, Rockefeller explained his motivation as his "love for, and gratitude to, the city which has always shown me kindness . . . and . . . has afforded me much pleasure to make this contribution to her welfare and prosperity." With Rockefeller's contribution the city could now boast of owning an entire river—Doan Brook—as an idyllic replacement for the fallen Cuyahoga.[31]

Such gifts were not born out of regret for the environmental havoc created by industry. J. G. W. Cowles, Rockefeller's personal real estate agent, delivered a full-throated defense of the industrial development of Cleveland at an annual meeting of the Cleveland Chamber of Commerce in the spring of 1894. "The Cleveland of 1796 was a wilderness, with no mark of civil order," Cowles began by reaffirming the organizing power of industry to tame the frontier. Although he did nod to the natural advantages of Cleveland, Cowles emphasized entrepreneurial efforts to transform the environment. "[T]he land and the waters were here, also the skies and the forests; but that was all." Perhaps, the thinking seemed to run, smoke and acid represented the price of progress. "The skies were blue, the air clear and pure, the grass green and the trees abundant and luxurious," Cowles continued. "No odor of oil or acid, and no black pall and stain of coal smoke offended the senses of the dwellers in the Forest City." Cowles affirmed the dominant ethos that guided the business of the era by stating "What Cleveland is, has come to pass by favor of nature and by force of individual enterprise." Responsible for Rockefeller's real estate interests in Cleveland, Cowles failed to mention how he had capitalized on these environmental changes by developing suburban communities that offered the middle class an escape from urban pollution.[32]

The city commenced a vigorous development campaign for the new park corridor. "Keep off the Grass" signs and an army of groundskeepers conveyed an overbearing and decidedly affluent vision for the new parks. Citizens soon discovered that thirty thousand dollars of municipal taxes had been earmarked for park police to patrol suburban parkland far beyond Cleveland's city limits. A critical editor of the *Cleveland Recorder* noted, "numerous policemen [are] kept away out in the barren country about the old Shaker place, where no one except an occassional [*sic*] squirrel hunter ever goes, and where the park commission is spending hundreds of thousands of dollars, improving an allotment for Mr. John D. Rockefeller, at public expense." It became apparent that the new parks were intended for the class of people who came to the parks to view their aesthetic grandeur rather than working people who simply needed a place to stretch their legs and get lost in the foliage after a day spent in the acrid Cleveland air.[33]

Within a year of Rockefeller's gift in 1896, public opposition to the park commission culminated in the formation of the Park Board Reorganization Association (PBRA), which sought to make the parks accessible to the urban poor. The association won the sympathy of the urban population by drawing attention to how the new parks, while drawing their revenue from the general city fund, raised the property value of the rich who (with their tax dollars) fled urban pollution to suburban enclaves. Several vocal citizens argued that the creation of the parks along Doan Brook represented a boondoggle for real estate settlements financed by the same "philanthropic" donors to the park system. John Zangerle, a lawyer and leader of the critical PBRA, discovered that nearly all the recent park donations given to the city came with the stipulation that the park board expend funds, precise figures often stipulated, to improve the land and maintain it in perpetuity. If the city failed to live up to these conditions, the land would be forfeit to its original owners. In gifting the city 278 acres of land covered in ponds and swamp, for example, the Shaker Heights Land Company inscribed considerable control over how the land would develop. Paragraph 5 of the deed states the "grantor, its successors, and assigns, for the period of ninety-nine years from the date of this conveyance [April 10, 1896], shall have the right and privilege to open streets through the abutting lands and connecting with said park drives, such streets to be opened and connected at such intervals and in such manner not to impair the usefulness of said park driveways." The following paragraph ensured the creation of this infrastructure by demanding the expenditure of no less than $150,000 within six years. In essence, the strip of suburban parks used the legal power of the deed to siphon urban taxes into suburban pleasure grounds.[34]

The park board, appointed by a sinking fund commission that itself served indefinitely, drew these funds from the municipal taxes of Cleveland, whose borders ended at the manicured lawns of the suburbs. The connection between public gift and private gain was often striking, as when it was revealed that Rockefeller's beneficent gift of $50,000 worth of parkland allowed his real estate bonds in the neighboring Euclid Heights allotment company to appreciate in value of $1,000,000. Philanthropy had become a wise business decision. The lawyers that headed the PBRA uncovered the economic and political horse-trading that made such schemes possible. L. E. Holden, a mining capitalist and president of the *Cleveland Plain Dealer*, seems to have enjoyed creating conflicts among his various interests. As a member of the park board, Holden directed city funds to improve land he either owned personally or increased the property value of nearby tracts in which he had a personal interest. Holden owned property surrounding Western Reserve University (where he also served as trustee), which linked Rockefeller's gifted land to the Shaker parks to the south. Intended as an escape from Gilded Age industry, Cleveland's parks also represented an extension of business into an emerging landscape of health—the suburbs.[35]

Toward the end of 1897 the independent Cleveland media—the *Leader* and *Plain Dealer* were controlled by Rockefeller and H. E. Holden, respectively—began to trace the money trail between the new parklands, local banks, and Cleveland's park board. The *Cleveland Recorder* discovered that many people associated with the park board, through either service or philanthropy, held positions at the Union National Bank of Cleveland. The clerk for Cleveland's park board held the same position at the bank before his appointment by the city. The Cleveland Park commissioner's brother served as a director for the Union National, whose vaults held park board funds. Andrew White, a member of the sinking fund commission that appointed the park board, simultaneously served as attorney for the Union National, thus helping to ferry his own associates from the bank to the park board. It is fitting that the famous Cleveland Republican Mark Hanna headed the Union National Bank. Recently appointed to the US senate, Hanna summed up the extent to which the economy and politics had merged by the 1890s. "There are two things that are important in politics," the famed Republican suggested. "The first is money and I can't remember what the second one is." In Cleveland, wealth also determined the geography of environmental health.[36]

The opponents of the Cleveland Park Board turned their attacks toward the deeded land and the antidemocratic nature of the board itself. To ensure the expenditure of the stipulated amounts required in the deed and to reward partisan friends, the park board spent exorbitant sums, which caught the attention of its critics. Shortly after the centennial announcement of the park gift, the *Cleveland Recorder* argued, "hundreds of thousands of dollars have been worse than wasted. . . . In many instances, the law has been openly violated by not advertising for the lowest bids in the letting of contracts, and in almost all the small things the most expensive, instead of the most economical, methods have been employed." The *Cleveland Press* focused on the "Nearly $150,000 spent for park [b]ridges which are characterized as being an offense to good taste and grossly extravagant [*sic*]," citing the opinions of "leading Cleveland engineers" who denounced "the 24 bridges . . . as being inartistic and failures." The criticism could very well have been legitimate or a form of professional jealousy. The park board hired a Boston engineer and landscape architect, Ernest W. Bowditch, at a salary of twelve hundred dollars per month to oversee the work normally done by local engineers for as little as three dollars an hour. The board already had forty-two engineers on its payroll at the time, which further raised suspicions of intentional waste and mismanagement of public funds.[37]

During the winter of 1897–1898 the press received further condemnation of the park board by its own employees. Edward Pfaff, a park policeman who patrolled the 276-acre Rockefeller gift, confessed to the *Cleveland Press* "that there were 60 engineers and assistants kept on during the entire winter." The *Press* discovered these "engineers and assistants spent their time in playing pedro . . . [and] had a pair of boxing gloves and would spar and shot at a mark to kill time."[38] Opposition to the park board reached new heights as reports of waste and abuse leaked out. By February 1898, the PBRA boasted seven thousand members, and no fewer than eighty-two clubs within the city had, in the previous year, "declared against the present management of the park system and passed resolutions asking for a change in the management." Even leaders of Cleveland's religious communities voiced their concern at the abuse of charity to enrich the affluent. In December 1897 Moses J. Gries, rabbi of the Temple at Central Avenue and East Fifty-Fifth Street, addressed the Cleveland Chamber of Congress and argued that "the park commissioners ought to be responsible to the people." Gries challenged the emerging geography of health by stating "parks should not be four or five miles away from the houses of the poorest, but just across the street."[39]

Charles E. Bolton, soon-to-be mayor of suburban East Cleveland, initially attempted to negotiate affordable access to the new parks through favorable street rail rates. When the compromise was rejected in early 1897, Bolton resorted to attacking Cleveland's park board as an antidemocratic tool of the wealthy. "We, the citizens of the East End, do heartily favor swift, cheap, direct, and commodious transportation." Bolton concluded by couching his demands in the iconic Progressive phrase, "The greatest good to the greatest number in Cleveland and country will be accomplished by completing from Erie street to Case avenue the electric railway."[40]

As public apathy eroded and community leaders organized an articulate critique of the park board, its advocates in the press and chamber of commerce sought in a last-ditch effort to stave off democratic reform. In February 1898, the *Cleveland Leader* warned of mob rule if the "present board will be destroyed and in its place will be created a board or a single director under the absolute control of the mayor." The editor believed elections for the board "will be used for the strengthening of the City Hall machine, and the big contracts for park and boulevard improvements can be used for the promotion of political interests rather than the welfare of the city and its people." Instead, the *Leader* sought to characterize the present park board as constituted by successful men motivated by noblesse oblige. "We believe the members of the present Park Board have been inclined to do what was fair and just."[41] The *Leader* was suggesting that democracy should not extend to Cleveland's public spaces, a sentiment echoed in a letter to the editor submitted under the pen name Columbiana: "It is not every man that possesses the ability to spend large sums of money in the way that will get the most of it." Rooted in the new geography of health created by air pollution, the battle over Cleveland's parks had now raised the question of whether democracy was still relevant in an era of big business.[42]

Mayor Robert McKisson responded to the *Leader* in an editorial appearing in the *Cleveland World* by arguing that democratizing the appointment of the park board "will not increase my power politically. It will do just the reverse. Anyone who knows anything about politics will tell you that the greater the number of appointments a man has to make the more enemies he will have." On the same day the PBRA took the opportunity to paint the defenders of the status quo as antithetical to the founding beliefs of the American system of government. "The park board must go," the PBRA urged. "It is a system of taxation without representation, and a removal of the whole business from the people and placing of it in the hands of those who are as far removed from the taxpayers as possible." The antidemocratic nature of the park board became the strongest argument the PBRA communicated to the press. The PBRA revealed that a "careful inquiry in 44 of the most important cities in the United States shows that everywhere . . . park commissioners are either elected by the people or appointed by officials or bodies that are elected by the people." Furthermore, the PBRA suggested the board itself might well have been "either wittingly or unwittingly, . . . a party to a huge real estate deal in so far that it agreed to put through the vast improvements" stipulated by the land deeds it accepted throughout the decade.[43]

By early 1898 advocates of the park board made a final attempt to preserve the status quo by falling back on the rhetoric that throughout the Gilded Age had served them in defending business from government regulation. In a Machiavellian move that would make the Koch brothers proud, they invented support for their position where none had existed before. James H. Hoyt, a prominent Cleveland lawyer and general counsel for the Cleveland Chamber of Commerce, helped organize the Non-Partisan Park Board Association to attack the claims of the PBRA and give the appearance of public support for the present board. At the first meeting of the board, he asked, "Would you, who are richer in this world's goods than I, be willing to make any donation to the park system of Cleveland if it were managed by those who are willing to engage in city politics? I deny that the Park Board is un-American. It is good business. Nothing is un-American that is good business."[44]

NO WALK IN THE PARK

The defense came too late. The PBRA hammered the final nail in the coffin of the park board by revealing the disproportionate use of funds by the sinking commission. The parks located at the eastern extremity of the city had garnered $1,707,352 in city funds by the beginning of 1898, while all other city parks received only $583,437. After residents from the city's West Side petitioned the park board for modest improvements to Edgewater Park west of downtown and were disregarded, the die had been cast for the majority of the city to overthrow what they saw as a tool of East Side real estate interests. A. G. Daykin, president of the Western Improvement Association, argued that the parks "were given to the citizens of Cleveland and not merely to the citizens of one section of the city."[45]

Never as monolithic as opponents viewed it, the business community splintered on the issues of parks and clean air. An anonymous writer to the *Recorder* in February 1898, argued for a reassessment of priorities. "It is simply disgraceful the condition that some of our business streets are allowed to remain in," the citizen contended, "while money is being spent by the hundred thousand in contracting pleasure grounds and pleasure drives for the rich of our city." Businesses downwind from Standard's refineries betrayed little sympathy for the notion that industry should operate free of restraints. The abstract of the suit brought against Standard Oil Company by William Macey describes the perils of associating industrial smoke with prosperity. Macey operated two tenements on a rise overlooking the debouchment of the Kingsbury Run into the Cuyahoga. In the decade after he purchased the property, the Standard Oil Company would fill the land between Macey's tract and the river with the largest oil refinery in the world, Standard's No. 1 Works and its forest of smokestacks. Macey claimed the "noisome exhalations, acid fumes and vapors of the products of coal tar, caustic and sulphuric acid and other substances in countless degrees of combinations" filled the air and settled on his property. First, the trees perished. Then the vegetation shriveled and died. His tenants moved out, and his family grew sick and suffered from sleeplessness. Claiming "he was prevented from the enjoyment of his property," Macey sued for four thousand dollars in damages in the Cleveland Court of Common Pleas. As in so many other cases, Standard's lawyers settled the case out of court, thereby avoiding legal precedent and publicity.[46]

The early history of air pollution in Cleveland serves as a reminder of the conflicts at the core of the early conservation movement. As Cleveland's working communities darkened from air pollution, the PBRA argued that citizens, regardless of their rank or station, deserved access to healthy environments as a fundamental right. "Parks are for the enjoyment of the people," the PBRA declared in a press bulletin at the height of the battle over Cleveland's park board. "There should be no restrictions upon the enjoyment of the very things for which people go to the parks, namely, wandering at will through the fields and forests and over the grass and lawns." Yet the PBRA noted that the goals of increasing adjacent real estate values ran counter to these simple pleasures. "The large police force and numerous 'Keep off the grass' signs serve to make a visit to the Cleveland parks one of anxiety rather than pleasure," concluding that "even now, of the 1,200 acres of pleasure grounds the people may eat their lunches in [only] three spots selected by the board." Such elite forms of conservation proved so unpopular that by 1900 the park commission had been abolished, and the city embarked upon reform in government that characterized the nationwide Progressive Movement. With the election of a populist mayor, Tom Johnson, a year later, the "Keep off the Grass" signs came down and the new Division of Parks and Boulevards dedicated much of its time to building urban playgrounds and installing ball fields over the landscaped grounds of their suburban parks.[47]

CONCLUSION

Thomas Gavagan, a native of Ireland, gives modern city dwellers a view into the life of the poor neighborhoods in the industrial core of Gilded Age Cleveland where the majority of the population lived. Pushed into a meeting of the board of health by friends and neighbors, Gavagan complained of the industry that robbed him of his peace. "[T]he shtame and the stink is so bad I have to close me kitchen windys. The machinery shakes the house and make the childer sick an' I can't slape a wink when I worruk nights." It is important to remember that Gavagan's experience with air pollution represented but a small snapshot of a corporation that operated refineries throughout the country by the 1890s. The relationships forged between industry, environment, and community repeated themselves in similar fashion anywhere the petroleum mode of production converted crude oil into kerosene. Following the construction of a massive refinery in Bayonne, New Jersey, residents of Staten Island voiced familiar concerns about the Standard Oil Company. In 1908 the New York state commissioner of public health reported that Standard's operations threatened "the comfort, repose, health and safety of a considerable number of persons in the County of Richmond," forcing Standard to install smoke abatement scrubbers in their towers.[48]

By the beginning of the twentieth century, little had changed for residents within Cleveland. Even the office of the smoke inspector had shriveled in the face of several court challenges. During the 1890s, as capitalists in the Forest City utilized the power of the chamber of commerce to beat back a surge in labor unionism, local judges challenged the authority of the city government's meager attempts to regulate business. In 1895 a judge declared Cleveland's smoke ordinance unconstitutional, forcing the city to respond by winning passage of a bill in the state legislature "granting to municipalities the power to declare what is a nuisance and giving it power to abate the same." Without it, not only would the city be unable to act against polluters but the police force might find itself on shaky legal ground in enforcing a range of laws. Two years later, again the smoke ordinance was thrown out in court when Judge Fielder ruled, "declaring that it was a prohibitory ordinance and that the State conferred no such powers on the City Council, but gave them power to regulate and not to prohibit." The mission of Cleveland's Office of the Smoke Inspector could change overnight on the interpretation of a single verb by a county judge.[49]

When John Krause assumed the position of chief inspector in 1902, decades of business resistance and court challenges forced him to focus on "education rather than coercion." He spent his tenure in the smoke inspector's office delivering public lectures, meeting with smoke inspectors from throughout the country, and attempting to work with business. His actions did little for the atmosphere he was charged to protect. The editor of the *Cleveland Tribune* lashed out against Krause for "patting himself on the back too much and allowing the city's air to remain smoky."[50] In the end, the smoke inspector was only as good as the legal climate supporting him and many decision-makers in Cleveland viewed its air and water quality as an unfortunate but necessary sacrifice for material wealth. "Cleveland is a manufacturing center," Dr. Martin Freidrich, a contemporary of Krause and a fellow health officer, stated in 1902. "The smoke that annoys us has made the city what it is, the metropolis of Ohio." The frontier ideology associating the domination of nature with progress died hard in Gilded Age America.[51]

However the affluent tried to escape the toxic air and water of the urban core, the wind eventually caught up with them. In 1904, despite all the efforts by Rockefeller and after dismissing warnings from public health officers (including the chief forester of the US Department of Agriculture), Cleveland approached a grim milestone in its environmental history. In his annual report to the city council, the city forester reported a mass death of trees within Gordon Park—the lakeside jewel of the five-mile emerald ribbon of parkland. Decades of struggle to enforce even modest abatement measures led the forester to the conclusion that "a continuance [of smoke] means the entire destruction of the nice bit of natural woodland, which is the last fragment of native forest in this locality inside the city limits." The city of Cleveland, which entered the Gilded Age with an almost embarrassing amount of woodland, emerged from a quarter century of industrial growth unable to bequeath to the twentieth century a single tree that stood when Moses Cleaveland disembarked with his crew of surveyors at the mouth of the Cuyahoga and founded the city in 1796.[52]

FIVE

EFFICIENT EARTH

The final chapter of this story shows how the Standard Oil Company constructed a new landscape that fully embodied the lessons of industrial efficiency by extending control over the four classic elements as well as the surrounding community. Creating a company town allowed Standard Oil to extend the Gilded Age partnership between business and government well into the twentieth century. This "long" Gilded Age came to end in the 1960s when the environmental consequences of the efficiency game spilled out of Standard Oil's corporate fiefdom and captured the attention of a national audience, now demanding federal intervention. Although the bulk of historiography on the Standard Oil Company focuses on the realignment of capital that was pioneered by the 1882 trust agreement, the Gilded Age's premier petroleum corporation initiated an equally novel reorganization of the environment in the closing decades of the nineteenth century. The consolidation of Standard's business empire prior to the 1880s was predicated on three goals: acquiring property, refining technology, and eliminating competition.

The creeping realization of existing environmental limitations—particularly the declining control it exerted over the diminishing Pennsylvania oil fields and the restraints of its power over the politics and environment in its Cleveland home—led the corporation to a new, almost totalitarian, perspective toward its relationship with labor, politics, and the environment. This realignment emerged on the southern frontier of suburban Chicago, amid the sand dune marshes of Lake Michigan's Indiana shoreline where the company constructed a massive refinery, which is still in operation today.

By the mid-1880s, Rockefeller's Standard Works No. 1 in Cleveland no longer shipped its products to the Atlantic coasts. Recently annexed refineries in Brooklyn, Bayonne, and Philadelphia now served the Atlantic market. Instead, the Cleveland Works, once the largest in the world and Rockefeller's stepping stone to market domination, was reoriented to serve only the domestic market. Much of the refining capacity of Cleveland's works was transferred to a refinery rising among the Indiana sand dunes on the southern shore of Lake Michigan. In many ways, the move was predetermined by natural limits. Rockefeller had already consolidated all the refining properties in Cleveland, most of which bordered his own massive No. 1 Works on the steep banks of the Kingsbury Run. Cleveland's industrial flats had long been filled by steel mills, lumberyards, and other manufactories, leaving only marginal lands left to expand upon. The petroleum industry also learned its first lesson in the finitude of oil. In 1869 the Petroleum Producers' Association was organized in the Pennsylvania fields to limit production in the hopes to both boost crude prices and postpone the eventual decline of the field. Crude production in the western Pennsylvania oil fields leveled off in the late 1880s and peaked at just over thirty-six million barrels in 1900. Unlike whales or trees, oil does not grow back, making every field a once-in-history bonanza. Nature's limits turned Rockefeller's gaze to new opportunities.[1]

Although Standard controlled 92 percent of the petroleum flowing from the Pennsylvania oil fields as late as 1880, a disorganized producers' union prevented the company from having direct control over the source of crude. A new field in the vicinity of Lima, Ohio, offered Standard the promise of complete control of production that was impossible in Pennsylvania. Few competitors thought to enter the Lima-Indiana field because the high sulfur content of its oil produced foul odors when burned. In 1886 Standard plucked an inventive young chemist, Herman Frasch, from a competing petroleum business, and a year later Frasch filed papers with the US Patent Office for his cleansing process, providing the oil trust with the technology to unlock the commercial potential of the vast Lima-Indiana field and maneuver around the Pennsylvania producers' union.[2]

By 1886 Standard had organized a pipeline company to connect the Lima-Indiana field to Chicago. Wishing to avoid another crowded industrial city, however, Standard leaders set their sights on the lakeshore of northwest Indiana for a manufacturing heart to the new Lima-Indiana field. What Rockefeller desperately needed in order to achieve his dreams of scientific efficiency were large tracts of contiguous flat land oriented between new producing fields, transportation networks, and blossoming continental markets. The corporation originally considered construction of the new refinery in South Chicago, which served as the terminal for its Lima pipeline, but the lack of land, the communal opposition to pollution, and the prospect of falling into the orbit of Chicago's political machine forced it to reconsider.[3]

The Calumet region of northwest Indiana proved ideal. Although the *Chicago Tribune* described the location as a "sandy waste," the lake dunes and marshes represented a rich array of microclimates. Even a Standard official, never shy about pointing out nature's need for improvement, called the environment of the region a "veritable wonderland" of biological diversity.[4] Inhabited at various times by species more suited to swampland, desert, or forest, the dune lands of southern Lake Michigan harbored over thirteen hundred species of plants including ample grape vines, strawberries, cranberries, wild rice, and over three hundred varieties of birds. Before Standard Oil came to the Calumet region, affluent Chicagoans were drawn to the open air and hunting prospects so close to the city. They rented quarters from local farmers and built several grandiose lodges in the fertile country. Chicago also drained the Calumet region of raw materials, drawing it into market relations that prefigured its transformation into an industrial suburb. Chicago firms scoured the region for timber to rebuild from the devastating fire of 1871, and Armour operated an ice house in Lake County, Indiana, employing several hundred men who worked throughout the night with the aid of electric lights to supply the company's refrigerator cars of dressed beef. This single operation on Red Cedar Lake shipped as many as sixty train carloads of ice per day back to Chicago, significantly lowering the lake's water level. Industry had drawn the natural resources of the Calumet region into commodity markets before Standard Oil arrived in 1889, but these transformations did not significantly alter the foundation of the region's ecology—the kaleidoscope of ecotones between lake, marsh, forest, and dune settings that maintained a rich biological community.[5]

In 1889 the veteran superintendent of Standard Oil's Cleveland works,

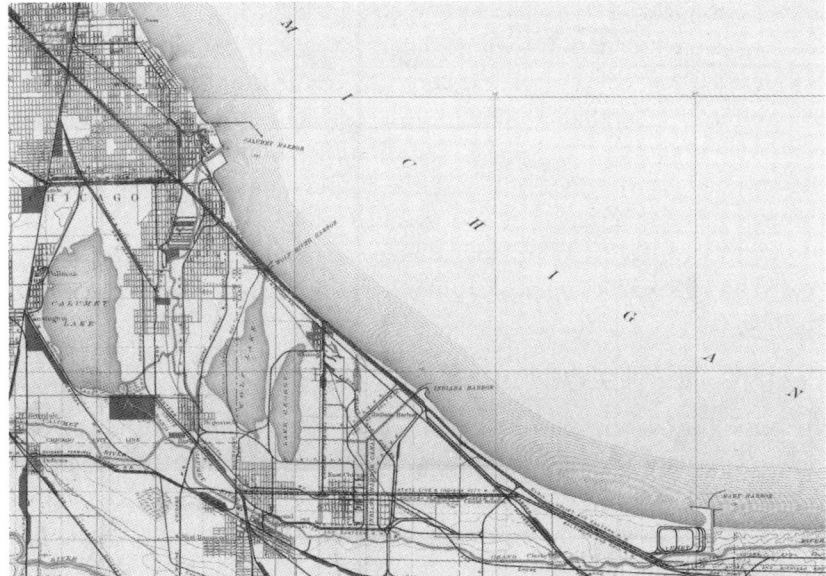

After the construction of the Indiana harbor and canal just east of the Illinois-Indiana border, the Calumet region enjoyed economies of scale in connecting its industrial capabilities to resources and markets throughout the Great Lakes.

Credit: "Chicago Waterways, Sheet No. 2, Map Showing Transportation Railway Conditions," U.S. Engineer Office (1911), Map Collection, University of Chicago Library.

William P. Cowan, surveyed the lakeshore three miles east of the Illinois-Indiana state line near the quiet post office of Whitings Siding and initiated the secretive purchase of over three hundred acres of land, with the help of local general store owner Henry Schrage. Cowan also organized the initial work crews of experienced laborers from its refineries in the East. On May 5, 1889, laborers began the work of leveling sand dunes and filling marshes. During construction, workers required boats to travel about the marsh-strewn storage tanks, but Standard's engineers lowered the water table through the construction of an elaborate sewer system and the surrounding wetlands became level, desiccated property for the company. With plenty of land at their disposal, Standard spaced their massive storage tanks five hundred feet apart, a luxury impossible in the crowded Cuyahoga valley. Workers constructed sand levies around the storage tanks, initially to help drain the land, but later reinforced them to serve as spill reservoirs to contain the threat of fire. From the beginning Standard Oil sought to safeguard profits by setting the terms of their relationship with the environment.[6]

Perhaps the largest natural obstacles to the company's plans of achieving a level, rationalized landscape were the many lakes and swamps that dotted the Calumet region. One of the larger lakes, Berry, stood in the way of Standard's dreams of control. The company constructed an elaborate sewer system for its emerging refinery to deliver waste water to Lake Michigan, effectively dooming Berry Lake. The location of the most desired hunting grounds prior to the coming of the corporation, Berry Lake splashed through the new sewer pipes and reunited with Lake Michigan. By 1891, however, Berry Lake had become a mere memory, as the local population waded into the remaining pockets of water with buckets to haul back a final fish harvest.[7]

SEEING LIKE A CORPORATION

To supply the massive refinery and the growing community of workers with water, in January 1890 the corporation sank a five-foot-diameter tunnel under the bed of Lake Berry and extended it a half mile out into Lake Michigan. With the completion of the sewer system and the waterworks tunnel, the Calumet region's hydrology had been remade to suit the industrial needs of Standard Oil. By 1891 the Whiting refinery connected to Standard's headquarters at 26 Broadway in New York City via telegraph wire and began operations, converting over five million barrels of crude into petroleum products a year. By 1895 Standard Oil encouraged its workforce to vote for incorporation of the city of Whiting so as to merge the political interests of its workers to their paymaster. Standard managers, including William Cowan, observed the vote in the summer of 1895, which resulted in only two souls out of 687 objecting to the corporation's plans. When Whiting conducted its first election the following September, the local newspaper, the *Whiting Democrat*, endorsed Standard's candidates, proclaiming, "Everyone who knows anything about the Standard Oil Company knows that it does what it does well and economically. It follows, therefore, that the town will be well and economically conducted." The election resulted in the superintendent of the refinery, W. S. Rheem, winning a seat on the board of trustees, and then earning quick promotion to president. Within a decade, company officials effectively controlled the government from the board of trustees up to the town's first mayor, William E. Warwick.[8]

Of the first seven mayors of Whiting spanning the years 1903–1954, four held management positions at Standard Oil. Two of the first three mayors—William E. Warwick and Beaumont Parks—served as superintendents of the Whiting refinery while in office. Of the three mayors not employed by the corporation, two (Fred J. Smith and Walter E. Schrage) held top positions at First National Bank of Whiting, which had been organized by Henry

Schrage with the funds acquired by collaborating with Standard Oil during the initial purchase of land in Whiting. Within a decade the corporation had succeeded in reengineering the Calumet lakeshore and in setting the terms of the social contract between local government and its citizens. By the close of the nineteenth century, a visiting journalist from Chicago could declare without exaggeration, "Whiting is the Standard Oil [C]ompany embodied in a town." The new century would witness an intensification of these social and environmental relationships as new industries migrated to the Calumet region, attracted by the possibility of establishing their own corporate fiefdoms.[9]

In 1901 the Lake Michigan Land Company agreed to deed fifty acres east of Standard's refinery to the Inland Steel Company on the condition that the corporation spend no less than one million dollars in constructing their new works. Additionally, the land company agreed to build a harbor within a year, capable of berthing the largest ore vessels on the Great Lakes. The Calumet region, already covered in railroad trunk lines connecting Chicago to the East, would now add a steel mill to Standard's massive refinery—and a harbor where nature had neglected to provide one. The Lake Michigan Land Company acted out of self-interest in the deal. Formed in 1895, the company dreamed of cashing in on their real estate in the Whiting area by constructing a harbor linked via canal to the Grand Calumet River. Inland Steel's 1901 commitment brought the plan to fruition when Indiana Harbor opened in 1903.[10]

Like Whiting, the harbor and steel mill transformed what had once been a backwater into an industrial city overnight. A mere "railway switch tower in July, 1901," one reporter noted, "today [December 1902] it has 350 houses . . . 2,000 inhabitants, and the deepest harbor of any city along the whole south shore of Lake Michigan." Although blowing sand still collected in waves on the newly paved streets, Inland Steel followed Standard Oil's lead by employing 75 men to grade land and another 250 to build a sewer system. The city of East Chicago that grew up around the steel mill and harbor grew fivefold in the first decade of the twentieth century, with immigrants comprising half the population.[11]

Andrew Carnegie's steel empire also followed Rockefeller into the Calumet region. Soon after organizing the U.S. Steel Company in 1901 by joining resources with J. P. Morgan, Carnegie built a massive mill opposite Inland Steel at Indiana Harbor for a sum of two million dollars. The mill employed 800 men and paid three hundred thousand dollars for riparian rights to both the harbor and the Calumet canal, still under construction. In 1906 U.S. Steel began constructing not only the largest steel mill in the world but

Workers for US Steel moving the Calumet River on July 21, 1906.

Credit: CRA-42-100-002, folder 2, box 100, US Steel Gary Works Photograph Collection, 1906–71 (CRA).

an entire city just east of Whiting and Indiana Harbor near the source of the Grand Calumet River. Gary, Indiana, named after the founding chairman of U.S. Steel, Albert H. Gary, extended the industrial transformation of the Calumet region that had begun at Whiting a full ten miles from the Illinois-Indiana border. In constructing the massive Gary Works, U.S. Steel moved the Grand Calumet River a half mile to the south and straightened the riverbanks with steel bulkheads and slag from its mill. It cut yet another canal from this new channel and created the region's second artificial harbor so as to allow the mill access to lake traffic. The corporation also forced the Calumet to flow less like a swamp and more like a true river by expelling several million gallons of wastewater into its new bed every day.[12]

The histories of these industries interlink in important ways. In December 1902, Inland Steel entered a contract with Standard Oil to purchase fifty thousand gallons of fuel oil per week. By 1907 a pipeline delivered fuel directly from refinery to mill, where Inland had converted their operations almost exclusively to petroleum fuel oil and had constructed a fifty-thousand-gallon-capacity tank. John D. Rockefeller was partly responsible

REFINING NATURE

for the coming of the steel mills to the region. During the economic panics of the 1870s, fellow Clevelander and steel baron Samuel Mather convinced Rockefeller to buy up the ore-bearing Mesabi Range in northern Minnesota. Northern Indiana proved an ideal middle ground to manufacture the raw material into commodities that were then easily distributed by ship or rail to the entire continent. In its first full year of operation, Inland drove a modest sixty-two thousand dollars' worth of sales. In the following years they amassed enough capital to secure a lease to their own mine in the Mesabi Range in 1906 and to form the Inland Steamship Company in 1911, achieving an impressive vertical integration in only ten years of operation. In 1917, at the height of the First World War, Inland Steel exemplified the material growth of the whole Calumet region, producing over a million tons of ingots and clearing just over fifty million dollars in net sales, a spectacular eight-hundred-fold increase from its first year of operation less than a generation prior.[13]

Standard Oil's errand into the wilderness proved fabulously successful. As early as 1896 the *Chicago Tribune* noted how the "manufacturer has recognized in the Calumet region a location after his own heart." Beginning with a few hundred acres of swamp and sand, the company constructed the largest refinery in the world, capable of converting 2,500,000 barrels of crude oil into a spectrum of commodities each year, amounting to one-tenth of all refining production in the United States. With this success, however, came the ecological consequences of concentrating a significant proportion of the nation's refining and milling capacity along the shore of the fourth-largest freshwater lake in the world.[14]

WARNING SIGNS

Chicago—like Cleveland, Standard Oil's birthplace—was a city defined by the interplay between people, a central river, and a massive freshwater lake. Chicago discovered this early as material success, especially in the meatpacking industry, proved inversely related to the health of Lake Michigan, the primary source of drinking water after the Civil War. According to historian William Cronon, Chicago became the "gateway to the Great West" in the nineteenth century. In his exhaustive environmental history of the city, *Nature's Metropolis*, Cronon argued that Chicago's empire began to unravel in the closing decade of the nineteenth century. The city had grown too big, and its success bred imitation by upstart western towns. The metropolis served as the core of a massive hinterland, stretching across the plains to the peaks of the Sierra Nevada range. By the closing decade of the nineteenth century, Chicago faced severe environmental limits to its growth. The city's engineers successfully reversed the flow of the Chicago River between 1871 and 1900, redirecting tons of industrial and residential sewage from the source of its drinking water in Lake Michigan to the Des Plaines River and on down the Mississippi River. Just when the city seemed to secure the health of its water supply, the Calumet region emerged as an industrial dynamo and opened troubling new sources of pollution.[15]

The Grand Calumet River found its headwaters near the shore east of Gary, Indiana, but debouched into Lake Michigan on the Illinois side of the border. Before industry began to dredge and move the river, it meandered through wetlands. These Calumet marshes served as a dumping ground for Chicago's southern suburbs and industry in Whiting and nearby Hammond, Indiana. At the close of 1895 a Chicago firm charged with clearing the slough for navigation reported that, although 248,516 cubic yards of earth had been dredged from the river, the "work has been worse than useless, as the channels have been filled in by the deposits from slaughterhouses, manufacturing establishments, and sewers in the vicinity of Hammond." To make matters worse, refuse from Indiana threatened Chicago's water intake cribs in Lake Michigan, leading Congressman L. E. McGann to report that the "Calumet district is sending more sewage into the lake right now than did the Chicago River when the city had 750,000 inhabitants."[16]

The Calumet region quickly gained a nefarious reputation for its pollution. The *Chicago Tribune* in 1902 presented its readers with colorful directions to northwest Indiana. "In going," the paper reported, "you should begin to pick up your baggage just at the moment when through the car windows comes an overpowering smell of gasoline, petroleum, Vaseline, coal tar, kerosene, coal oil, paraffine, and other 187 smells that belong to the Standard Oil company; for that is Whiting, Ind., and Indiana Harbor is only a few yards beyond the most pungent and nauseating of these odors." The writer concluded, "Whiting is all right in its way, but most people will continue to prefer the Chicago stockyards." Chicago residents who had come to Whiting to hunt game or to escape the bustle of the city for a quiet afternoon picnic bristled at the loss of their refuge once Standard Oil arrived. As construction of Standard's refinery proceeded in the 1890s, riots broke out between Chicago campers and Standard Oil construction workers at the picnic grounds near the Lake Shore and Michigan Southern Railway station. The conflict ended only after the corporation's police officers arrived and superintendent Cowan secured the cessation of these train services. Chicago was losing control over its hinterland.[17]

By 1908 most changes to the Calumet region occurred on a narrow spit of land between the lakeshore and the Grand Calumet River. That year, the

As industry flocked to the Calumet River in the twentieth century, businesses created new real estate from fill. For example, this apron of land at the mouth of the Indiana harbor and canal (note the Y shape extending to the left edge) housed several blast furnaces operated by Inland Steel. Smoke rises from Standard's massive refinery (center top), which occupies most of the land between the Indiana Canal and the city of Whiting.

Credit: Calumet Regional Archives, Indiana University Northwest.

battle to make nature conform to productive industrial plans entered a new era when Randall W. Burns, who owned twelve hundred acres of marshland, began a decades-long effort to reclaim the Calumet marsh. A historian of the region, Powell A. Moore, compared the project to the famous draining of the Pontine Marshes in Italy. The canal system, known as Burns Ditch, redirected the Little Calumet River and some tributaries east of Gary at Ogden Dunes, the new frontier between industrial suburb and the dune lands of the Calumet region. The project created over twenty thousand acres of drained marshland that would soon become an apron of real estate for steel mills, farmland, and residential development. With the completion of Burns Ditch in the 1920s, the entire Lake Michigan shoreline from the Illinois border to Gary and extending inland to the Little Calumet River, in the words of a University of Chicago geographer, "is almost completely industrialized, except for a few very small parks." The Calumet region, in little more than a generation, was transformed from a diverse ecosystem supporting only a limited economy centered on hunting, fishing, and farming to one of America's most significant industrial hearths. As exploitable land disappeared, the corporations that remade the region turned to Lake Michigan for new land.[18]

The shifting sand that authored much of the landscape became a commodity early in the region's history and found use as fill for Chicago's Columbian Exposition in 1893. In 1898 as many as three hundred railroad cars of sand left the dune lands every day. As the dunes disintegrated, industrialists turned to their source—the lake. Lake vessels and steam "sandsuckers" scoured the lake bottom near the Calumet shore and sold the reclaimed land. Inland Steel began construction of a second blast furnace in 1906 but lacked adequate land between the ribbon of railroads and lakeshore. The company brought in a steam shovel from the Mesabi Range and employed it in expanding the shoreline into the lake. After eleven years, dozens of acres of land had been summoned from the lake bed at the cost of $750,000.[19]

Residents fought these actions by forming associations and challenging the legal basis for private interests to convert waters held in common by the state into a private commodity. Chicago won stringent protection of public riparian rights in the landmark US Supreme Court decision *Illinois Central Railroad Company v. Illinois* (1892), but with corporations firmly in control of Calumet politics, development of the lakeshore continued unabated on the Indiana side of the border. In 1907 the General Assembly of Indiana passed an act declaring "the owner . . . of land bordering upon the waters of Lake Michigan shall have the right to fill in, reclaim and own the submerged land adjacent to and within the width of his land . . . and may build docks, wharves and other structures thereupon for industrial . . . purposes." The lakeshore had now been legally severed from its ecological role and remade into property as sandsuckers and steam shovels belched out new real estate in the now crowded Calumet region.[20]

"A SPOON AT A TIME"

The Frasch Method allowed Standard Oil to solve the problem of high-sulfur Lima crude. But, the slurry of chemicals did more than cleanse petroleum. During distillation, hydrogen sulfide vapor formed, which was both toxic and a dangerous explosive hazard. Hydrogen sulfide acts against the body in a similar way to carbon monoxide, but in high enough concentrations it can kill a man with a single breath. The British tried to capitalize on this property and deployed it as a chemical agent against the Germans during the First World War. Standard's Acid Works became so perilous that the company resorted to the use of canaries as indicators of dangerous concentrations of the gas. The copper oxide treatment and the array of caustics used to remove sulfur from crude corroded both flesh and metal and even turned the hair and skin of workers and draft horses green. Wives forced their husbands to undress on the porch before entering homes, and families stocked up on woolen clothing because, as one former Standard employee noted, "wool clothes, wool pants would last about maybe a month and a half, two months. . . . [b]ut regular cotton pants they wouldn't last no more than two weeks. . . . [a]nd maybe only one day if you get sprayed with acid or something, it was gone."[21]

As early as 1899 botanists from the University of Chicago's Hull Botanical Laboratory discovered that the smoke and soot from the refinery had "injured or destroyed" the coniferous forests near Whiting. The nuisance became so troublesome that the company was forced to move an entire residential neighborhood of its own workers from the former location of Berry Lake. Refining high-sulfur Lima crude also produced prodigious amounts of petroleum coke residuum that the company was able to market to Whiting workers as fuel for their domestic coal boilers, and this further contributed to the community's air pollution. The mountains of petroleum coke the company could not sell were dumped into Lake Michigan at the company's pier near Whiting Beach. Some of it went to form a breakwater, although much of it was simply dumped in the open water, causing bathers to scrape their feet on the jagged lumps of residuum.

In 1906 journalists from Chicago reported on a region in environmental free-fall. A writer for the *Chicago Tribune* dubbed Whiting "The dead city of the dead." In interviewing a local housewife, the reporter asked, "Why don't you plant things in your front yard?" The woman responded, "Deh smoke," leading the reporter to conclude, "It would be a waste of time to try to grow things for beauty in that atmosphere. In spots there is grass, but it is smudged and blackened from the smoke." The historical record discloses a growing anxiety among the residents of the Calumet region as the transition from a subsistence ecology to market economy took the appearance of a one-way street. In writing the first company history, thirty years after the founding of the refinery, Whiting journalist U. G. Swartz captured the reaction of the local community to these vast environmental changes. "Those who had sold their places to the Company not only had to leave their homes and land," Swartz wrote, "but they had to stand by and see all that they cherished throughout demolished." This environmental dysphoria, Swartz reported, "was a sort of home-sickness which they could not get over. They preferred hunting and fishing and the genial tasks of their contented

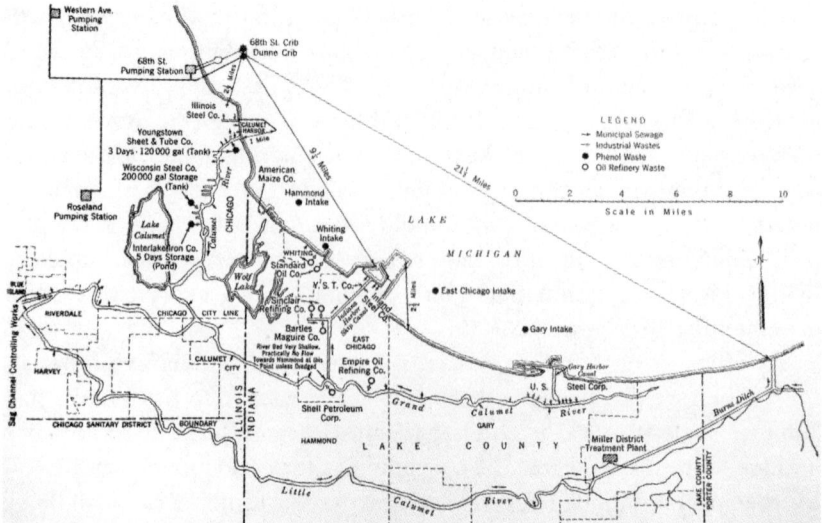

Following the outbreaks of typhoid fever in the 1920s, sanitary engineers investigated how industry located in Indiana jeopardized Chicago's water supply. This map, created in the early 1930s, illustrates the proximity of municipal water intake cribs to oil waste sources clustered around the Indiana harbor ship canal. Note the "made land" just east of Whiting absent in early maps of the Calumet.

Credit: Gorman, "Survey of Sources of Pollution."

lives to the paths which might lead to more active scenes and busier days and the fruits of a sturdier life." This sense of loss mirrored the concerns of the growing conservation movement during the Progressive Era.[22]

The region's posh hunting clubs withered and died, as drainage schemes dried up the surrounding wetlands. The Gary Country Club sold much of its land, which was made into a golf course. The Calumet Gun Club rented its hunting cottages out to engineers of U.S. Steel. One club, named after the Kankakee River at the southern border of the Calumet region, was purchased by a Chicago industrialist after quarry left its lands for greener or, in this case, damper pastures. William Cameron tried to manufacture his own game preserve by reflooding the 135-acre tract with irrigation ditches fed from the Kankakee River. Despite Cameron's quixotic obsession, the environmental trend for the Calumet district was toward a landscape of fungible, desiccated real estate that harbored far fewer plant and animal species than the mosaic of dunes, marshes, and lakes it replaced.[23]

Although skilled workers imported by the company completed the initial construction of the refinery and residential areas, once the plant opened Standard Oil sought new immigrants (particularly eastern European Slavs) to serve as a low-cost labor force. This new immigrant labor force was more familiar with the landscape quickly vanishing than the industrial world of wage labor taking shape on the outskirts of Chicago. Many were drawn to the few remaining common lands in an effort to supplement family income through hunting, fishing, and foraging. Children of the immigrants in particular stood at the crossroads between nature's economy and the new industrial order. One Whiting resident recalled the "hunting paradise" of Wolf and George Lakes, where he would trap muskrat and sell skins to Sears Roebuck. "And it was nothing for us to see kids going out with shotguns under their arm," Joseph Novosel Sr. recalled, "When the duck season began, at seven o'clock in the morning the Standard Oil whistle would blow, and all the shooting would start out there."[24] By the time historians from the University of Indiana caught up with Novosel in the autumn of 1990, he pessimistically reflected on the century of changes that had converted Whiting to an industrial zone, concluding "that's what happened to our community, we had a lot of these people that came in here, made a fast buck, polluted, tore up the place, but they didn't do it like overnight. They did it like a spoon at a time, so you never noticed it." Joseph Gresko, also interviewed as part of the University of Indiana's oral history project, offered a more blunt view of the environmental history of Whiting. "You know, like Standard Oil, OK, that's my bread and butter, still is today, they raped the city of Whiting. You take a look at our lakefront over there. Take a good look at that lakefront. That is a disaster area."[25]

The most significant transformations occurred in Lake Michigan. After the First World War, no fewer than four million people relied on large municipal waterworks tunnels tapping the lake. Oil pollution in Lake Michigan proved problematic because Chicago and its suburbs relied on chlorination to sanitize the public water supply. Oil and caustics reduced or neutralized chlorine's sanitizing effects. In 1923 this exacerbated a typhoid outbreak that struck 227 people, killing 23 Chicagoans. The following year, typhoid claimed the lives of 21 people in Hammond, Indiana, when lake pollution compromised chlorination efforts. In the following decade, the engineers of a single intake crib found that the effectiveness of chlorination had been reduced on 155 separate occasions due to petroleum contamination. Chicago's sanitary engineers identified the refinery at Whiting and water from the Indiana Ship Canal as the source of the waste and the Indiana legislature, at the urging of the state board of health, passed a bill prohibiting the pollution of any stream or lake. However, they included a special provision that exempted Lake Michigan and any streams flowing into it.[26]

Following the First World War, the water pollution problem precipitated a bi-state effort to engineer the waste away from the source of drinking water for the region. In 1922 engineers attempted to feed the Calumet River into the Chicago Sanitary and Ship Canal with the construction of the Calumet Sag Canal. The engineering effort faltered when rainstorms or low lake levels fed the river back into Lake Michigan. Despite these engineering efforts, water quality declined in the 1930s. When a yearlong US Public Health Service study linked lake pollution to industries in the Calumet, Chicago and the State of Illinois sued the cities of Hammond, Whiting, Gary, and East Chicago (as well as the state of Indiana) for compromising Lake Michigan. Four years later the case was dismissed by a Master in Chancery, who deemed the defendants had accomplished "certain corrective measures." The city of Chicago was hardly in a position to advocate for self-sacrifice to achieve a common goal. The state of Missouri sued the state of Illinois in 1900 following the diversion of Chicago's wastewater down the Mississippi. Other lake states sued the city in 1922 fearing Chicago's river diversion would cause lake levels to fall and endanger shipping. The US Supreme Court issued a decree in 1930 directing Chicago to cap the canal flow at fifteen hundred cubic feet per second, which it accomplished through the construction of a mammoth water treatment works in 1949. Reengineering nature to fit industrial needs created a domino-effect through the justice system, materialized in the form of costly sanitary works, and revealed the multiple stakeholders in the ecological stability of Lake Michigan.[27]

The Public Health Service studies compiled in the 1920s and 1930s suggested the coastal waters of southern Lake Michigan had passed a tipping point. "The pollution of Lake Michigan by sanitary sewage and industrial wastes, especially from Indiana," read one report, rendered the drinking water in Whiting "unfit for that purpose, even with elaborate and efficiently operated purification plants." The superintendent of the Whiting Water Department, in a paper published in the trade journal of the American Water Works Association in 1941, reported, "Since the time of that report, of course, conditions have become progressively worse." Whiting's intake crib lay at the mercy of surrounding industry, which had constructed a mile-long breakwater to protect Indiana Harbor and had pushed out the shore toward the crib with "filled in" land, bottling the "untreated sewage from an estimated population of 250,000 and the industrial sewage of steel mills, oil refineries . . . and innumerable miscellaneous manufacturing establishments" in an artificial bay. The city of Whiting responded by constructing an ozone treatment works in July 1940, the first of its kind in the United States.[28] Ozone occurs naturally from lightning, but harnessing this power industrially through a twelve-thousand-volt direct current, the superintendent of the works admitted, occurred "through action not thoroughly understood." The experimental works cut chlorine demand in half and lowered threshold odor numbers of the water to as low as eight, which was still nearly three times the limit of three set by the US Public Health Service. Despite this radical solution, Whiting's water quality continued to decline, reaching odor numbers as high as three thousand over the next generation.[29]

Water pollution ignored political boundaries, crossed into Chicago's coastal waters, and endangered that city's three primary intake cribs at mid-century. Chicago constructed two massive filtration plants: the South Water Filtration Plant (1947) and Central District Filtration Plant (1964). At the time of their completion, each became the largest water-processing facility in the world, allowing the city to supply three billion gallons of filtered water to its residents daily (a little less than what flows over Niagara Falls in a day). Pollution from Indiana Harbor and the Calumet River spiked at mid-century, challenging public infrastructure. Sanitary engineers monitored the movement of "slugs," patches of concentrated wastes drifting along lake currents. With southerly winds, some slugs made a circuit of all of Chicago's major intake cribs. At times they even reached the crib for the city of Waukegan, Illinois, fifty miles north of Indiana Harbor. These pollution slugs required Herculean efforts by engineers, who stood ready to feed alum coagulant, chlorine, and activated carbon into the water supply to protect water quality. By dumping wastes into public waterways, Calumet industries socialized costs to the entire region. Water company engineers calculated that every pound of ammonia waste that entered an intake crib required ten pounds of chlorine to maintain water quality. Inland Steel, US Steel's Gary Works, and Standard Oil (now American Oil Company) together discharged 34,350 pounds of ammonia each day into Lake Michigan and the Calumet River.[30]

CRASH

Whiting Beach, which had been one of the last remaining stretches of dune land in the community, was a rare communal meeting spot for fairs and became a popular destination for teenagers seeking refuge from the smoke and din of the town. Like most of the southern shore of Lake Michigan, shallow waters extend far out from the beach into the lake. By the middle of the twentieth century, the actions of the sandsuckers, the dumping of refuse slag from the refinery, slicks of ballast "slop oil" dumped from the company's lake vessel fleet, and the proximity to the company's sewer outlet combined to make what used to be a simple swim now into a nightmare. In 1910 Standard Oil's own Lighterage Department's handbook on "Rules and Regulations" demanded that any "waste oil" not burned under the boiler be "thrown overboard" to avoid fire hazards. The US Coast Guard reported twenty-eight oil spills from ships in Lake Michigan in 1967, with the majority still occurring in the Calumet region. Many residents complained of swimming in oil refuse that stained their skin and in floating human excrement dumped from the company's nearby sewer outlet. Others discovered the land had become so saturated with petroleum as to make the act of constructing a beach fire a harrowing experience. According to one resident, "when I was like ten, eleven, and I recall digging a hole, putting a match in that hole, and guess what? Catch on fire."[31]

Industry in the Calumet region was situated along one of the prime spawning areas at the shallows of southern Lake Michigan, which only magnified its damage to the ecosystem. Frank N. Egerton's research on lake ecology reveals both the importance and the vulnerability of highly productive shallows. Entire species of fish disappeared from the lake in the first fifty years of Standard's Whiting Works. Fishermen noted how high-value but environmentally sensitive species such as lake trout and whitefish fell from just over half of the annual catch at the turn of the century (about ten years after the construction of the refinery) before crashing to 4.4 percent of the harvest by the late 1950s. These changes had a significant impact on the regional economy in the first half of the twentieth century, for the Lake Michigan catch represented as much as half the total commercial fishing harvest in the entire United States at this time. Industrial pollution can cause both qualitative and quantitative changes to shallows ecology. In moderate amounts, according to Egerton, "pollution has merely resulted in the displacement of highly preferred species intolerant to pollution by less preferred species which can live with pollution." Once pollution drives oxygen levels below certain thresholds, however, the ecosystem crashes. Extensive studies performed in the wake of the 1948 Federal Water Pollution Control Act described an ecosystem in collapse. High concentrations of industrial and residential pollutants led to an "overfertilization" of Lake Michigan. Ammonia nitrogen and sewage fed naturally occurring phytoplankton populations, which exploded in numbers, died, and in places reduced dissolved oxygen to zero, wiping out the food chain. Off the shore of Chicago and the Calumet region, the lake bottom ecology was reduced to sludge-worms, aquatic scuds, bloodworms, and fingernail clams. Indiana Harbor and the Grand Calumet River consistently registered dissolved oxygen levels of zero.[32] Engineers described the Calumet's river system and coastal waters as "barren biologically," containing little more than "rubble, petroleum wastes,

REFINING NATURE

and a heavy black oily organic ooze that had a highly objectionable sewage and petroleum odor."[33]

Although untangling the variables causing these changes is complex—if not impossible—for a system as large as Lake Michigan's 22,400 square miles, the changes wrought by Calumet industries certainly favored some species over others. For example, the parasitic sea lamprey prefers warm streams for spawning. By stripping northwest Indiana of its forest cover, leveling and draining its sustaining wetlands, and burdening the waterways with decomposing effluent, Standard Oil and the industries that followed it into the Calumet region helped to raise water temperatures and began an avalanche throughout the entire lake ecosystem. When finally measured in the middle of the twentieth century, the former wetland landscape of surrounding Lake County boasted the highest erosion rate of any county bordering the Great Lakes, losing 8.3 tons of soil per acre every year. Lampreys preyed on high-value lake trout and whitefish that together accounted for nearly 40 percent of the catch in the early 1940s when the lamprey began to garner attention. A decade later, lake whitefish accounted for less than half a percent of the catch, and trout had entirely disappeared from the commercial market. The alewife, which had followed the sea lamprey from the Atlantic coast to the Great Lakes, now free of predation from the whitefish and trout, took advantage of these changes during this same period. Not even present in the commercial records as late as the 1950s, the alewife exploded in population until they represented nearly two-thirds of the commercial harvest in the mid-1960s, having essentially replaced the now maladapted species.[34]

Nature delivered the coup de grace for the stability of Lake Michigan in the form of a record four-year drought from 1962 to 1965. By the second year of the drought, water levels in Lake Michigan fell to their lowest recorded levels in eleven out of twelve months, dropping lake levels by as much as one foot four inches in the spring of 1964 and exacerbating existing pollution concentrations and temperature spikes. In the summer of 1967 the increased nutrient load dumped from residential and industrial sources caused alewife numbers to peak as the fish feasted on a booming zooplankton population. On June 15, an official from the Federal Water Pollution Control Administration patrolling sources of water pollution from his US Navy hydroplane noticed curious white lines in Lake Michigan. As the plane dipped lower, the official made out the unmistakable windrows of dead alewives. Americans have recorded alewife die-offs in the Great Lakes dating back to the nineteenth century and as early as the eighteenth century in the Atlantic Ocean, but human memory and recorded history failed to prepare observers for the scale of the unfolding environmental catastrophe that July. As the hydroplane made a grim tour of the crisis, the official witnessed "one great shimmering band of alewives stretching for 40 miles between Muskegon and South Haven, Michigan."[35] Through a combination of disease, lack of food, terminal oxygen levels, and temperature shock, the alewife suffered a Malthusian crash and died in spectacular fashion. The Federal Water Pollution Control Administration estimated a total of twenty billion alewives washed up along the shores of Lake Michigan in the subsequent weeks. Alewife carcasses blanketed the beaches, clogged water intakes, and manifested the contradictions of industrial efficiency. Chicago's Park District sprayed five thousand gallons of deodorant onto its beaches; Michigan's governor, George Romney, pledged eight hundred thousand dollars for a "cleanup program"; the federal government passed a bill devoting millions to fund federal oversight; and *Time* magazine brought the nation's attention to the collapse in its nascent "ecology" section.[36]

When asked by reporters if the problem warranted millions, "if not billions," of federal tax dollars, Murray Stein of the Federal Water Pollution Control Administration replied soberly, "We can't afford not to do it—because if we don't, Lake Michigan may be lost forever."[37] The environmental crisis that began on a local level in the Calumet region in the 1890s and spread to Chicago in the twentieth century had captured national attention and served as a set piece of the blossoming environmental movement. A statement from Chicago's Hyde Park–Kenwood Community Conference —a grassroots environmental organization—put it simply, "Most of this waste is poured into Indiana waters, but the lake recognizes no boundary lines; the waste pours into our beaches and into our drinking water. Only Federal action can stop the gradual ruin of Lake Michigan."[38]

Although some residents in northern Indiana quietly complained about the changes wrought to the region's ecology by Standard Oil, the company could count on social discipline from its workers. It paid 67 percent of the taxes in Whiting, and company executives dominated the city government well into the twentieth century. Most Whiting residents interviewed following the community's centennial agreed with local historian Betty Gehrke who said, "everybody noticed the odors but we'd always say it was our 'bread and butter,' 'don't knock it.'" For nearly a century, the owners and management of the Whiting refinery avoided both government regulation and public anger because the devil's bargain of economic prosperity at the cost of environmental health allowed them to implement their vision of the ideal city. As early as 1906 a reporter for the *Chicago Tribune* marveled at the control the company exercised over Whiting, writing, "every soul in the place

Chicago Park District employee Philip Saeli sprays deodorant on dead alewives at Montrose harbor in July 1967.

Credit: Bob Langer, photographer. Courtesy of Sun-Times Media.

looks to the Standard Oil company for bread and butter, or cake and ice cream, or wealth, just as the case may be." It seems little had changed in the intervening decades.[39]

But there were competing measures of wealth as well. Representing 4.5 percent of the world supply of freshwater, Lake Michigan had been since the retreat of the glaciers what one marine biologist called a "conservative system." Seventy years of industrial change remade it into an aquatic reflection of the boom/bust cycle of consumer capitalism. Industrialism also made it home to millions of Americans. The price of economic growth and increased efficiency was the degradation of common resources that cities such as Chicago and Hammond were then required to expend significant public funds to protect. Chicago's two record-shattering filtration plants for example, when completed in 1964, cost taxpayers $460 million. Flirting with the limits of efficiency cost a fortune.[40]

As Craig Colten has argued, factories of the Calumet are more than mere "relics of an extinct breed . . . benign reminders of a prosperous era." They mark a "massive graveyard" where the cornucopian dreams of the Industrial Revolution lie buried. As industry scaled down and fled the region, the history of the Calumet echoes that of the larger Rust Belt and looks little different from the great boom towns of the American West. The promise of wealth never seems to offset the environmental and social dislocations produced in our rush to convert nature into commodities. By 1965 a report commissioned by the Federal Department of Health, Education, and Welfare warned that the ecological change to Lake Michigan was already "practically irreversible," even with "the cessation of present waste discharges." Seventy years of heavy industry had significantly altered a system that had been built by thousands of years of coevolution in the footprint of the Ice Age.[41]

Before he became the first administrator for the newly created Environmental Protection Agency in 1970, William Ruckelshaus served as counsel to the Indiana State Board of Health. In the early 1960s he helped draft the first air pollution law in the state's history and pursued cases of "gross pollution." He quickly learned the limits of government regulation. The "competition among the states for the location of industry within their borders," Ruckleshaus discovered, created a race to the bottom to attract business. "But whenever we pushed a major company very hard, there was always the threat they would move to the south where the governors said, in effect, 'Come on down here, we don't care, we need your business, we need jobs.'" Public servants, from the smoke inspectors in nineteenth-century Cleveland through to the modern EPA, have discovered that the market will always find a path around legal barriers to nature's wealth.[42]

CONCLUSION

In 1915, after years of promising graduate study, Alice Mabel Gray withdrew from her college courses and, with only a few provisions, a blanket, and two rifles, boarded a train in Chicago and headed to the dwindling dune country east of Gary. She was in many ways the product of Rockefeller's success. Gray used her undergraduate degree from the University of Chicago (founded by John D. Rockefeller) to acquire a position as one of the first women "computers" at the US Naval Observatory in Washington, DC, before her desire for further study brought her back to Chicago to pursue a graduate degree. After four years, her dreams soured, however. Although eager journalists theorized that her desire to flee civilization for the dune country emerged from a disappointing love affair, it appears her studies led her to adopt a naturalistic perspective that alienated her from her peers. For Gray, Chicago was the "child of Lake Michigan in a more poetic sense," and its greatest treasure was not its manufacturing establishments but, rather, its "rich black soil." After setting herself up in an abandoned cabin, Alice Gray became a curious specter in the dune country, bathing in the lake and sustaining herself on the abundant waterfowl she could bring down with her legendary aim and through bartering with local families for what the dunes failed to provide. After a year, the Chicago press heard rumors of a "nymph" that had abandoned civilization for the vanishing wilderness of the lake country and gave her the moniker "Diana of the Dunes," in reference to the Roman goddess of the hunt. If Rockefeller's goal was to drag Americans into an industrial future, Alice Gray was confidently marching in the opposite direction.[43]

The Chicago press sought out Gray and offered her an opportunity to voice her antimodern perspective. For Alice, the contrast between nature's economy and that created by culture had proved too much. Gray argued that her work in Chicago at best produced "little of importance" and at worst amounted to "a prostitution of my powers."[44] In short, Alice's perspective was the absolute obverse of Rockefeller's: "The life of a salary earner in the cities is slavery, a constant fight for the means of living," and she angrily lashed out at Chicago's residents for doing nothing to prevent the dune country from "being Garyized." According to environmental historian Andrew Isenberg, Chicago had become a "neocolonial power over the West" in the nineteenth century. But the story of Alice Gray suggests that the metropolis could cultivate a new cooperative relationship with the land.

In the end Gray's romantic idealism ended in tragedy. Although she befriended the local housewives and fishermen, even entering a romantic relationship with a transient furniture maker, Alice's decision to reject the wage economy and a traditional gender role led to rumors of thievery and immoral activities. When locals discovered a murdered corpse near her home in 1922, Gray and her shady beau became immediate suspects. A confrontation with a drunken sheriff's deputy left Alice with a cracked skull and her boyfriend, Paul, shot in the leg. Although the couple recovered from the incident and were released from police custody, they returned to find that their cabin had been ransacked. Three years later, exhausted by fighting a losing court battle against the Chicago press for libel, Gray died in Gary, the industrial cancer eating away her beloved dunes. Her last wish, to be cremated and her remains spread from the top of Mount Tom, a two-hundred-foot-tall dune in the heart of her wild sanctuary, was spurned by her family who trumped her wishes and demanded she be interred in Gary, the city she grew to hate.[45]

By 1923 with the completion of Dunes Highway, the new icon of the petroleum age had already penetrated the heart of dune country. Gary traffic officers marveled as over twenty thousand vehicles made their way from Chicago to the dune lands in the opening weekend, marking the full transition of experiencing nature as a source of sustenance for the mind and body to a mediated consumer phenomenon, safely absorbed from the comfort of a touring sedan fueled by petroleum. By then the Lima-Indiana field accounted for only 2.3 percent of total domestic output of crude, down from its high of just over 34 percent in 1899. Now the great midcontinent fields and those in California supplied over 75 percent of the industry's needs, and as the petroleum frontier moved into new regions the environmental consequences of refining moved with it.[46]

A RIVER BURNS THROUGH IT

By the time of Chicago's Columbian Exposition in 1893, Standard Oil had successfully developed new products to complement its sale of kerosene. In the 1880s Vaseline—a high-quality petroleum grease—earned a popular market as a cure-all for cuts, burns, and chapped skin. Paraffin replaced tallow in the manufacture of candles, matches, and chewing gum. Various chemical distillates of petroleum were used in the manufacture of paints, dyes, varnishes, paint removers, and astringents—replacing plant-derived products like turpentine. By the turn of the century, Standard's products illuminated buildings, greased train axels, cleansed and dyed fabrics, and protected baby butts from diaper rash. The corporation's products had insinuated themselves into every nook of the American market. The company's aggressive agents were able to give away lamps and stoves, knowing the monopoly's hold on the market left consumers with few options on how to fuel them.[1]

Many of these products were not produced by Standard Oil itself but by a growing number of affiliate companies that were controlled by Standard under its trust system. Created on January 22, 1882, the Standard Oil Trust Agreement created a centralized body composed of forty-four trustees, who effectively owned the entirety of Standard's corporate stock and also shares of its ownership in dozens of other companies. The trust form was essential to circumventing state laws—and Standard's original Ohio charter of incorporation—that prohibited private corporations from owning another company. A year prior, John D. Rockefeller's closest business associate, Henry Flagler, had instructed the company's chief counsel, S. C. T. Dodd, to inquire into "the feasibility of organizing a corporation for the purpose of holding stocks of corporations in various states and making the business of such local corporations subsidiary to the business of the Central Corporation." Dodd's response cited "the difficulty of obtaining a charter which confers the necessary powers . . . [i]t being against the policy of the law for one corporation to hold or deal in the stocks of another." Dodd noted the "great power which a Legislature has over corporations of its creation" and suggested the trust form would insulate the company from taxes levied on out-of-state properties. Standard incorporated under the state's laws, using the trust agreement as a legal firewall. The trust agreement created two new companies, Standard Oil of New Jersey and Standard Oil of New York, the latter of which became the base of the corporation's organization. With the signing of the trust agreement, Standard Oil's headquarters effectively transferred from Cleveland, Ohio, to 26 Broadway in Manhattan.[2]

In 1889 the state legislature of New Jersey, wishing to attract business, amended the state's constitution to provide unprecedented concessions to corporations. The new law allowed businesses incorporated in New Jersey to own companies operating in separate states, set the maximum franchise tax at one-tenth of one percent of issued stock, required only one company director to hold residence in the state, freed stockholders from liability for corporate debts, and waived the responsibility of the company to issue annual reports. Most significantly, the 1889 legal changes and subsequent acts passed in the following decade allowed New Jersey corporations to act as holding companies, effectively legalizing Standard's secretive trust form. By the turn of the century, Standard Oil and a host of America's largest trusts—including the much larger U.S. Steel Corporation—officially transferred their headquarters to New Jersey addresses. This was one of the largest flights of capital in world history. At the dawn of the twentieth century, New Jersey Standard openly owned shares in forty-one companies. The changes in the law allowed the state of New Jersey to fund its government entirely through corporate taxes.[3]

Many of the companies within the trust specialized in distributing Standard's products to available markets. By 1878 Standard Oil opened a sales office in San Francisco, which marketed oil north to the border with Canada, south to the border of Mexico, and west to the Hawaiian kingdom. Indeed, foreign markets accounted for no less than 50 percent of Standard's annual kerosene sales. Although the company had initially dealt with export merchants in coastal entrepôts like New York in order to deliver products to foreign markets, by the late 1870s Standard began eliminating or controlling such intermediaries. Standard bought interests in Meissner, Ackermann and Company, a consortium of export merchants, and then invested heavily in a refinery in Austria-Hungary. It owned majority shares of significant German and Italian distributing firms and began buying up shares of a Danish firm with an eye toward expanding its share of the Scandinavian market in 1888. By the 1893 Columbian Exposition, Standard Oil's agents were pushing their products in Montreal, Manchester, Frankfurt, Paris, London, Barcelona, Hamburg, Rio de Janeiro, Buenos Aires, Milan, Constantinople, Beirut, and Bombay. Standard even opened offices in Yokohama, Japan, despite the country's protective attitude toward foreign trade. By 1900 Standard had captured one-third of the petroleum market in Japan and operated a truly global trade network.[4]

Despite its success in entering global markets, new challenges met the company. In the spring of 1879, Standard's chief agent for Asia, William H. Libby, wrote from the Hawaiian kingdom to their New York firm, Charles Pratt and Company, that a "recently enacted statute prohibits importation [of oil] under 100° flash." This rare internal memo revealed Standard's attitude toward what it saw as a barrier to its products. Libby arranged an audience with the island nation's minister of the Interior to discuss the matter and even met with the police marshal who oversaw inspection "to whom I have endeavored to give a few 'points' which apparently have been well received." The memo reveals that Standard officers understood that its products might fail such tests, but the company could succeed by appealing to the interests of a few officials.[5]

When Henry Demarest Lloyd published his critical history of the company, *Wealth Against Commonwealth*, in 1883, Standard's public image had hit rock bottom. Some consumers, frustrated with government inaction, organized against what they viewed as predatory policies. Lloyd's book includes the history of two community boycotts of Standard's marketing efforts. The US Congress, pressed by the public uproar created by Lloyd's work, passed the Interstate Commerce Act in 1887 to ban railroad corporations from practicing rate discrimination and established the Interstate Commerce Commission (ICC) to investigate business and enforce the equal protection clause of the Fourteenth Amendment. Three years later Congress passed the Sherman Antitrust Act to criminalize monopolies found in restraint of free trade. From the passage of the ICC, American laissez-faire attitudes waned as the federal government took on an increasing role in the economy, placing Lloyd's work among a select few books that precipitated immediate government action.[6]

Changes in the economy, technology, and environment delivered long-term challenges to the company. The age of kerosene, which built Standard to imposing heights, would slowly end. The development of electricity—beginning with Thomas Edison's first demonstration of the incandescent lightbulb at Menlo Park, New York, on New Year's Eve, 1879—ushered in a new regime of illumination that proved far more efficient, more cost-effective, and much safer than kerosene. Electricity's high initial capital investments, however, would give kerosene a half-century window before closing it for good. Although the electric light required ample coal power and carried its own consequences for nature and society, economically it proved far more convenient than kerosene lamps, which stank, required constant refilling and cleaning, and could easily become incendiary bombs. Kerosene lamps persisted well into the twentieth century, but the product no longer served as the lifeblood to Standard Oil's survival. After 1903 kerosene sales entered a period of permanent decline as the electric light devoured Standard's primary market, effectively ending the age of kerosene's dominance.[7]

Standard suffered other setbacks at the turn of the century. Although it had obliterated its competition in the previous generation, the earth then revealed new sources of petroleum to competitors. By the 1880s oil fields near Baku on the Caspian Sea came online with heavy investment from the Rothschild and Nobel financial dynasties. Refined in Fiume, on the Adriatic coast, Russian oil immediately posed a challenge to Standard's overseas sales. It was cheaper to produce and distribute domestically than Standard's products, and the Rothschild and Nobel financial empires successfully petitioned European governments to levy a tariff on American petroleum so as to support continental production. Historians Ralph Hidy and Muriel Hidy estimate that Standard held a virtual monopoly on the world petroleum trade in 1882. With the help of a heavy tariff in 1887, however, Russian oil had captured 22 percent of the world market by 1888. Three years later, Russian oil represented 29 percent of the world trade, eating into Standard's foreign sales.[8]

The opposition to Standard multiplied with every passing year. Rockefeller, wracked by health problems, withdrew from his company in the 1890s and handed the reins to his inner circle, just as his oil empire faced dire new threats. Standard's hold over the dwindling kerosene trade took its biggest hit ten days into the twentieth century, when a well situated atop Spindletop Hill just south of Beaumont, Texas, drilled into a pocket of pressurized crude oil, causing a spectacular gusher. Overnight, the well tripled American oil production, and it would lead to the creation of a cadre of domestic competitors, including Gulf Oil and Texaco. It appeared the sun had set on Rockefeller's empire.[9]

Following decades of litigation, investigations, and congressional acts, the US Supreme Court dissolved Standard Oil into thirty-four independent companies in 1911. Writing for the majority, Chief Justice Edward Douglass White cited the Sherman Act of 1890 as the legal foundation for this radical action. According to Justice White, Standard Oil "took its birth in a purpose unlawfully to acquire wealth by oppressing the public and destroying the just rights of others." The opinion vindicated the muckraking journalists and competitors who warned of a company fueled by greed and corruption. "[I]ts entire career," Justice White explained, "is marked with constant proofs of wrong inflicted upon the public and is strewn with the wrecks resulting from crushing out without regard to law the individual rights of others." The words also confirmed the sentiments of Standard's victims, from the horribly burned Peter Golden to William Macey, who had suffered "sickness and loss of sleep" from years of life downwind from Standard's Cleveland refineries. The dissolution of the Standard Oil Company commenced an era of reform in American politics that would see the creation of both the federal income tax by constitutional amendment and the federal reserve system two years later in 1913.[10]

Although it reaffirmed the power of law over business, the decision had not fundamentally altered the corporate landscape. The Supreme Court would serve as final arbiter between good and bad trusts, but bigness, in itself, did not die with the 1911 decision. In his history of the Progressive Age, Richard Hofstadter wrote, "big business that grew big through superior efficiency was good; only one that grew big by circumventing honest competition was bad." Standard Oil faced an organizational but not an existential crisis.[11]

OLD WINE, NEW BOTTLES

The "baby" Standards weathered the Progressive reforms. The centralized control over information and organization that had united them at first now dispersed and metastasized among the new companies with ease. With the stock shares among the severed kin of Standard evenly distributed to its investors, the conservative bookkeeping of Rockefeller's empire had underestimated the value of its capital. Less than a year after the court action, share prices in most of the severed companies had doubled. Rockefeller, who had refused to sell any of his shares in the company he helped birth, saw his personal wealth skyrocket to $900,000,000, making him the richest man in history. On the eve of the Great Depression, the "children" of the Standard Trust, which was valued at just under a billion dollars prior to the 1911 breakup, represented four of the top ten wealthiest American corporations. By 1929 those four companies (Standard Oil of New Jersey, Indiana, New York, and California) held $3,930,000,000 in assets, an astounding 13.79 percent of all corporate wealth in the country.[12]

As the century marched on, these four corporations would remain among the ten largest businesses in the United States despite world wars, economic booms and busts, and even changes in their surface identity. To claim their independence from their tarnished mother company, the new corporations adopted new logos and monikers. The New Jersey branch became Esso (a thinly veiled reference to "S.O." for Standard Oil), and then Exxon. The New York company, became Socony (for Standard Oil Company, New York), and then Mobil. Standard's Indiana branch became Amoco (for American Oil Company), which it acquired in the 1920s. Standard Oil of California renamed itself Chevron following a merger with former Rockefeller rival Gulf Oil, which had grown into an oil giant in its own right on the Spindletop gusher. Although the twentieth century began with the dissolution of famous trusts, it would witness far more corporate mergers than anti-trust dissolutions, including the merger of Exxon and Mobil, Standard's largest children, in 1998. When the merger finally passed all legal hurdles in 1999, not only did the company become the largest corporation in the world but its very existence nullified the spirit of the 1911 Supreme Court decision. Despite economic downturns and the growth in power of the federal government, the corporate form became a permanent force in the twentieth century.[13]

The Supreme Court's 1911 decision fractured the company along many preexisting divisions. Although some of the upstarts became briefly dependent on their siblings for different levels of production, the mother company had succeeded in imprinting the gospel of scientific efficiency and vertical integration among its splintered limbs. Those companies lacking a critical level of production between well and filling station—such as the Indiana branch that could no longer charge the Whiting refinery with its own oil— were nonetheless staffed with veterans of the kerosene age. These men had learned the lessons of both vertical integration and scientific efficiency; many were former competitors of Rockefeller who had been annexed into the oil empire. John Archbold, a fiery critic of Rockefeller in his youth, eventually sold out to Standard and served as the company's president from the time Rockefeller retired until the 1911 Supreme Court decision.

Shunning nearly every opportunity to make public statements, both in courts and with the press, Rockefeller reversed his policy with the publication of his memoirs *Random Reminiscences of Men and Events* in 1909. The light, conversational writing style of his autobiography offered a blunt, but personal defense of Standard Oil's business practices. It gave voice to the founder of the corporation, who for decades had allowed his enemies to define him and the corporation he founded. The Rockefeller family sought expert counsel following the massacre of twenty-one miners, women, and children at Ludlow, Colorado, when the National Guard and officials of mining corporations controlled by John D. Rockefeller fired on striking workers. The family hired Ivy Lee, the founder of the modern public relations industry. Lee's advice helped rehabilitate the image of John Jr., and in the aftermath of the national tragedy, the family heir, on the advice of Lee, urged his father to mount his own public relations offensive. The elder Rockefeller gave his reluctant assent and authorized the writing of a historical biography in 1915. Lee contacted William O. Inglis, a reporter for the *New York Evening World*, and invited him to interview the patriarch and write a biography. Lee quelled concerns among the secretive family by reassuring Rockefeller, "you can be sure that anything he writes will be absolutely friendly." Over three years (1917–1920), Inglis intermittently lived with Rockefeller and compiled over fifteen hundred pages of notes and transcripts from his almost daily interviews. However, when Inglis completed the first volume of his history in 1923, a three-journalist panel selected by John Jr. to review the piece before publication—which included Ida Tarbell—panned the piece as hagiography and it never went to the presses. Inglis clearly enjoyed his life as Rockefeller's personal historian. He earned over eighty thousand dollars for his efforts, and he lived up to Lee's promise. In a letter to Rockefeller on his eighty-sixth birthday, Inglis celebrated the oil baron's "achievement in freeing the peoples of the world from ignorance, from suffering and from death." The Rockefellers wanted a sympathetic pen, but needed a biography that would pass muster with its critics.[14]

With Ivy Lee's guidance, the family's philanthropic adventures became featured in newsprint along with humanizing articles on the elder Rockefeller's golf game and participation in his local church. Lee was a master of persuasion, occasionally permitting access to the patriarch to friendly reporters. Rockefeller's famed penchant for handing out dimes to children, derided as a cheap stunt by his opponents, smacks of Ivy Lee's influence. The family's quest for a sympathetic history continued in the last years of Rockefeller's life. John Jr. and John D. Rockefeller III organized the search for a viable historian and approached Allan Nevins, DeWitt Clinton Professor of History at Columbia University, to deliver a scholarly defense of the company and the pater familias. After initial meetings, John Jr. wrote his father in 1935, assuring the patriarch, "We all feel [Nevins] will do a good piece of work and are glad at last to have the matter under way." Nevins, given unfettered access to the Rockefeller Archives and the Inglis interviews, published a pair of two-volume histories dedicated to the corporation and its founder—*John D. Rockefeller: The Heroic Age of American Enterprise* (1940) and *Study In Power: John D. Rockefeller, Industrialist and Philanthropist* (1953). His efforts, while met by some historians as an annotated apologia, represented the first legitimate histories to utilize sources internal to the company and established the efficiency narrative later championed by Alfred Chandler Jr. to describe Standard's rise to power.[15]

The Standard family of corporations soon adopted this practice of open, aggressive public relations. From an objective perspective, many of the changes Standard Oil publicized materially improved both working conditions and environmental quality around its properties. The "baby" Standards founded Refinery Loss Committees and assigned special engineers as "loss coordinators" to reduce the frequency and amount of oil lost through spills and evaporation. The company guidelines require loss coordinators to sample effluent streams, "[i]nspect lines and tankage for leaks," and "[d]iscuss problems and methods of reducing losses with operating personnel." Although the economic incentive behind such reforms made them inevitable, these changes served as important examples of the greater Standard empire taking responsibility for an aspect of its environmental impact. The companies popularized these reforms through "motion pictures, posters, and painted signs" as much to educate its workforce as to assuage public concerns.[16] Standard also pioneered the use of white paint on storage tanks after a study it commissioned revealed that "the use of white paint reduced gross evaporation from a tank by 20 per cent" and installed technological innovations in both its sewer systems and storage tanks—such as floating roofs and special flumes—that prevented, at a single refinery, the release of "40,000 to 60,000 barrels of hydrocarbons annually" into the soil and atmosphere.[17]

Standard companies also endeavored to instill an understanding of the corporation's public image in its employees. Mobil went so far as to issue a booklet to its traveling representatives entitled "Press Relations for the Mobil Traveler." In it, the company warned its employees "What *you* say will be taken by many readers to be what *the company* and *America* says. You can't get out of it by saying 'These are my own views and not those of my company.'" In order to not "embarrass you and damage the reputation and business relationships of the company," the booklet provided a list of appropriate behaviors and strict guidelines for interviews and casual conversations. The company took its public relations seriously, instructing management in another booklet to evaluate an employee's "poise," "tact," and to consider whether the employee was "a good member of the community in which we live" in personnel reviews. The company's desire for control was totalizing, enlisting its entire workforce in the construction of an image of the corporation as a responsible member of the community.[18]

LIMITS OF CONTROL

The biggest public relations coup, however, came with the philanthropic endeavors of the Rockefeller family. John Jr., who increasingly took control of the family's business and public image, became an ardent supporter of the growing conservation movement in the first half of the twentieth century. The heir to the Rockefeller family fortune provided public funds and real estate donations for the creation of Acadia, Great Smoky Mountains, and Shenandoah National Parks and was active in the fight to preserve Redwood groves in the West. While on a family jaunt to Yellowstone National Park in the summer of 1926, John Jr. met with the superintendent of Yellowstone and future director of the National Park Service, Horace Albright, who personally guided the family through northwestern Wyoming. Albright believed the park service should preserve the ecological integrity, as he understood it, of the land it protected. He took his entourage on a detour through the Snake River Plain south toward Jackson Hole, Wyoming, where he informed John Jr. of how commercial development and ranching threatened vital forage for Yellowstone's roaming wildlife. Rockefeller surprised his host by assenting and, using the tools familiar to the corporation, quietly established a proxy—the Snake River Land Company—to secretly amass nearly thirty-five thousand acres that were eventually incorporated into a greater Grand Teton National Park.[19]

By the middle of the twentieth century, the Rockefeller family had rehabilitated a tarnished public image and still maintained growing profits. Embedded within the Rockefeller family's environmental philanthropy was the logic that private charity served as a leveler for the ills of private wealth and industrialism. Philanthropy also served as a conservative response to government regulation and interference in the economy. Private interests created wealth and could clean up the mess they made, the logic seemed to run. As in many other instances over the company's history, the environmental legacy of Standard Oil came full circle along a river in the summer of 2011. On July 1, a pipeline operated by ExxonMobil burst, leaking sixty-three thousand gallons of crude oil into the Yellowstone River near Laurel, Montana. Of the many historical ironies involved, few seemed as bewildering as the fact that the spill came just as Senator John "Jay" D. Rockefeller IV (an heir to the Rockefeller family fortune) was advocating the passage of a pipeline safety bill. The Yellowstone River finds its headwaters in the Absaroka Range before flowing through the national park and plummeting over what is one of America's signature natural wonders, Yellowstone Falls at the Grand Canyon of the Yellowstone before meandering into southern Montana and into the Missouri River. Norman Maclean tapped into the American belief in the spiritual rebirth one can experience on those lonely stretches of Montana's ample watercourses in his best-selling *A River Runs through It*. Those same watercourses stirred up the braided legacies of efficiency in both its capitalist and conservationist forms as engineers scrambled to clean up yet another oil spill.[20]

The Gilded Age witnessed many limits to American power. Although business blossomed and technology offered the freedom from environmental limits, unprecedented urban poverty and pollution left most Americans with less freedom than their ancestors. Power for some begat dependence for others. The kerosene that illuminated the wilderness also darkened the skies and contaminated the drinking water of urban America. The persistence of petroleum disasters in the twenty-first century also reveals the illusory nature of our contrived dichotomies between economy and environment, industry and wilderness, even philanthropy and malfeasance. The legacy of Cleveland's role in the modern petroleum economy passed from memory as the social and ecological dislocations overwhelmed the city's coffers and corrupted an environment that had once been envied by the Founding Fathers. Instead, we remember the inferno on the Cuyahoga, a municipal government in default on its loans, and its infamy as the "Most Miserable City in America."

Nature, as the source of wealth and prime mover of the economy, unifies what human law and custom has attempted to sever into discrete parcels of property and social divisions of race, class, and gender. If there is a social lesson in the environmental history of petroleum, it is in the recognition that ecological problems are rooted in social problems. As illustrated in the history of Standard Oil, the sacrifice zones of modern industry emerge from the twin failures of government to defend the public good and of the market to contain costs within property boundaries. It has been a goal of this book to document how the failure to reevaluate social goals amid changing circumstances can undermine the original values of our predecessors, namely, prosperity, justice, and health.

The impulse of city governments to fix symptoms of social ills with technology persists to our day. Modern Americans share an optimism about the power of science to overcome natural limits. A Pew Research Center and *Smithsonian Magazine* poll published in 2010 discovered that most Americans believe the country might solve energy problems, cure cancer, and send humans to Mars by the year 2050. Scott Keeter, Pew's research director, argued, "If the U.S. has a national religion, the closest thing to it is faith in technology." Rockefeller's impulse toward efficiency appears to have colonized our own thinking about nature. Environmental historian Steven Stoll notes that modern Americans "have accepted efficiency as the soul of what it means to be green." Technology, for now, largely succeeds in alienating us from the consequences of industry, but the efficiency game still fails us at the points of contact between nature and our economy. Gravity still commands water to reconnect with the sea regardless of stock prices. Differences in temperature will continue to summon mighty gales irrespective of which political party controls Congress. And the sun will continue to power the biosphere long after the last drop of petroleum has been squeezed from stone. Nature disrupts our desire to refine it along social goals and in the process reminds us that it operates according to its own rules, which we ignore at our peril.[21]

NOTES

INTRODUCTION

1. "The Fortune 500," Time Inc., 2015, at http://fortune.com/fortune500/; "Business and Industry Sector Ratings," Gallup Inc., 2016, at http://www.gallup .com/poll/12748/business-industry-sector-ratings.aspx/. See the bibliography for just a fraction of the scholarly works concerning Standard Oil Company and John D. Rockefeller. This book owes a great debt to recent environmental histories on the petroleum industry by Brian Black, Christopher Jones, and Paul Sabin. I owe a particular debt to Andrew Hurley for the perspective of environmental justice he adopts in his fine essay "Creating Ecological Wastelands: Oil Pollution in New York City, 1870–1900," *Journal of Urban History* 20, no. 3 (May 1994): 340–64. I have strived to merge this scholarship with the excellent urban environmental histories by William Cronon, Matthew Klingle, Martin Melosi, Ted Steinberg, and Joel Tarr to tell the history of petroleum in urban landscapes.

2. The first quotation is from David Freeman Hawke, "John D. Rockefeller Interview, 1917–1920" (microfiche, Rockefeller Archive Center, 1984), 51. These records consist of transcripts of a series of interviews conducted between 1917 and 1920 by William O. Inglis, a reporter for the *New York World*. The interviews were granted for the purpose of producing an officially sanctioned biography of Rockefeller, but Inglis was unable to produce a publishable manuscript. The original interview notes, cited as the "Inglis Notes," are located in Record Group 3 at the Rockefeller Archive Center. They were later gathered in a microfiche compiled by Hawke, which has been published and is available outside of the RAC, for example, at the Kelvin Smith Library of Case Western Reserve University. The second quotation is from John D. Rockefeller, *Random Reminiscences of Men and Events* (New York: Doubleday, Page, 1909; hereafter cited as JDR, *Random Reminiscences*), 67. See also Greg Grandin, *Fordlandia: The Rise and Fall of Henry Ford's Forgotten Jungle City* (New York: Metropolitan Books, 2009).

3. Hawke, "Interview," 1291. See also Ralph W. Hidy and Muriel E. Hidy, *Pioneering in Big Business, 1882–1911* (New York: Harper and Brothers, 1955), 1:129; JDR, *Random Reminiscences*, 28–29; Leigh Phillips, *Austerity Ecology and the Collapse-Porn Addicts: A Defense of Growth, Progress, Industry and Stuff* (Croydon, UK: Zero Books, 2015); Donella Meadows, Jorgen Randers, and Dennis Meadows, *Limits to Growth: The Thirty-Year Update* (White River Junction, VT: Chelsea Green Publishing, 2004).

4. In his history of Coca-Cola, Bartow Elmore argues, "Coke's genius . . . was staying out of the business of making stuff." Bartow J. Elmore, *Citizen Coke: The Making of Coca-Cola Capitalism* (New York: W. W. Norton, 2015), 9. Timothy J. LeCain, *Mass Destruction: The Men and Giant Mines that Wired America and Scarred the Planet* (New Brunswick: Rutgers University Press, 2009), 7.

5. Phillips, *Austerity Ecology*; K. C. Clarke and Jeffrey J. Hemphill, "The Santa Barbara Oil Spill, A Retrospective," *Yearbook of the Association of Pacific Coast Geographers*, ed. Darrick Danta (Honolulu: University of Hawai'i Press, 2002), 64:157–62; Hugh S. Gorman, *Redefining Efficiency: Pollution Concerns, Regulatory Mechanisms, and Technological Change in the U.S. Petroleum Industry* (Akron: University of Akron Press, 2001).

6. Mike Polk Jr., "Hastily Made Cleveland Tourism Video," April 14, 2009, at https://www.youtube.com/watch?v=ysmLA5TqbIY/; "Cleveland Deemed Most Miserable City in USA," February 18, 2010, at http://www.reuters.com/article/us-cities-miserable-idUSTRE61H5WN20100218/; Kurt Badenhausen, "America's Most Miserable Cities," February 18, 2010, at http://www.forbes.com/2010/02/11/americas-most-miserable-cities-business-beltway-miserable-cities.html/.

7. Several studies have explored the relationship between rural farms and urban energy requirements. Hay farms, which emerged to feed the urban horse population, would disappear following the transition to gasoline powered vehicles. See Richard A. Wines, *Fertilizer in America: From Waste Recycling to Resource Exploitation* (Philadelphia: Temple University Press, 1985); Marc Linder and Lawrence S. Zacharias, *Of Cabbages and Kings County: Agriculture and the Formation of Modern Brooklyn* (Iowa City: University of Iowa Press, 1999).

8. The last dramatic burn, which captured the attention of the nation, took place in 1969. The first fire occurred as early as August 1868 when waste oil ignited on the river's surface near downtown, although the first major fires did not occur until the 1880s. See John Stark Bellamy, *The Killer in the Attic: And More True Tales of Crime and Disaster from Cleveland's Past* (Cleveland: Gray, 2002), 200; Brian Black, *Petrolia: The Landscape of America's First Oil Boom* (Baltimore: Johns Hopkins University Press, 2000).

9. *Cleveland Leader*, June 28, 1870, 2.

10. Grandin, *Fordlandia*; James C. Scott, *Seeing like a State: How Certain Schemes to Improve the Human Condition Have Failed* (New Haven: Yale University Press, 1998), 340. Andrew Isenberg discovered a similar story in *Mining California: An Ecological History* (New York: Hill and Wang, 2005). In California, public anger over the unfolding environmental disaster emerged as early as the 1870s. Standard Oil escaped these early political conflicts in Cleveland by creating the company town at Whiting, Indiana. By the 1960s, however, the environmental destruction reached across state lines and affected the drinking water of Chicago.

11. *Annual Report of the Trade and Commerce of Cleveland* (Cleveland: Cleveland Printing and Publishing, 1892), 115, The Greater Cleveland Growth Association Records (MS 3471), box 68, Western Reserve Historical Society, Cleveland, Ohio; see Alison Frank, "The Petroleum War of 1910: Standard Oil, Austria, and the Limits of the Multinational Corporation," *American Historical Review* 114, no. 1 (2009): fn 10.

12. Henry Demarest Lloyd, *Wealth Against Commonwealth* (New York: Harper and Brothers, 1894); Jonathan Adler, "Fables of the Cuyahoga: Reconstructing a History of Environmental Protection," *Fordham Environmental Law Journal* 14, no. 89 (2002): 93; Jonathan Adler, ed., *Ecology, Liberty and Property: A Free Market Environmental Reader* (Washington, DC: Competitive Enterprise Institute, 2000).

ONE | IMPROVED EARTH

1. Patricia Nelson Limerick, *The Legacy of Conquest: The Unbroken Past of the American West* (New York: W. W. Norton, 1987), 71, as quoted in Harland Prechel, *Big Business and the State: Historical Transitions and Corporate Transformation, 1880s–1990s* (Albany: State University of New York Press, 2000), 4. For Roy's take on Chandler's efficiency theory, see William G. Roy, *Socializing Capital: The Rise of the Large Industrial Corporation in America* (Princeton, NJ: Princeton University Press, 1997). The 1830s witnessed the rise of the great textile manufactories in cities such as Lowell, Massachusetts; see Theodore Steinberg, *Nature Incorporated: Industrialization and the Waters of New England* (Amherst: University of Massachusetts Press, 1991), 65.

2. JDR, *Random Reminiscences*, 65.

3. For wealth concentration statistics, see Alan Trachtenberg, *The Incorporation of America: Culture and Society in the Gilded Age* (New York: Hill and Wang, 1982), 4; "They Are the 5%," *The Economist*, November 1, 2011.

4. "The Journal of George Croghan (1750–1765)," in Reuben Gold Thwaites, ed., *Early Western Travels*, vol. 1 (Cleveland: Arthur H. Clark, 1904), 131–32, 133; "George Croghan's Journal, 1760–1761," in Thwaites, *Early Western Travels*, 1:107; Daniel K. Richter, *Facing East from Indian Country: A Native History of Early America* (Cambridge, MA: Harvard University Press, 2001), 168.

5. Robert A. Wheeler, ed., *Visions of the Western Reserve: Public and Private Documents of Northeastern Ohio, 1750–1860* (Columbus: Ohio State University Press, 2000), 15–17.

6. Although the impact of Bacon's Rebellion on colonial society has become a set piece of early colonial histories, far less attention is granted to the trans-Appalachian frontier following the Seven Years War and the establishment of the Proclamation Line in 1763. For the frontier's impact on the American Revolution,

see especially Richter, *Facing East*, 189–236; Colin G. Calloway, *The American Revolution in Indian Country: Crisis and Diversity in Native American Communities* (Cambridge: Cambridge University Press, 1995); Robert A. Wheeler, ed., *Visions of the Western Reserve: Public and Private Documents of Northeastern Ohio, 1750–1860* (Columbus: Ohio State University Press, 2000), 15–17.

7. Wheeler, *Visions of the Western Reserve*, 62, 109, 118–19. Historian Alan Taylor has established the effect that seemingly abundant natural resources had on settler attitudes toward the land during the early republic. See Alan Taylor, "'Wasty Ways': Stories of American Settlement," *Environmental History* 3, no. 3 (July 1998): 291–310.

8. Edmund H. Chapman, *Cleveland: Village to Metropolis* (Cleveland: The Press of Western Reserve University, 1964), 21. Population figures taken from Campbell Gibson, "Population of the 100 Largest Cities and Other Urban Places in the United States: 1790 to 1990," US Census Bureau (June 1998), at http://www.census .gov/population/www/documentation/twps0027/twps0027.html/.

9. "William Bullock's Journey from New Orleans to New York, 1827," in Reuben Gold Thwaites, ed., *Early Western Travels, 1748–1846*, vol. 19 (Cleveland: Arthur H. Clark, 1905), 151. Ted Steinberg has documented the various ecologic push and pull factors leading to the westward migration, see especially Ted Steinberg, *Down to Earth: Nature's Role in American History* (New York: Oxford University Press, 2002), 43–47.

10. Ron Chernow, *Titan: The Life of John D. Rockefeller, Sr.* (New York: Random House, 1998), 25 (FDR); Grant Segall, *John D. Rockefeller: Anointed with Oil* (New York: Oxford University Press, 2001), 15–16 (Bill). For the historical transformation of peddling into the sales profession, see Walter A. Friedman, *Birth of a Salesman: The Transformation of Selling in America* (Cambridge, MA: Harvard University Press, 2004).

11. JDR, *Random Reminiscences*, 28–29, 33. See also Chernow, *Titan*, 17, 22; "Interview, Mr. David Dennis (b. 1835)," Inglis Notes, folder 3-8, box 4, Rockefeller Archive Center, Rockefeller Family Archives. Sleepy Hollow, New York (hereafter RAC).

12. Inglis interview quoted in Chernow, *Titan*, 43–44. See also "Mr. G. W. Andrus interview," September 14, 1917, Inglis Notes, folder 9, box 4, Record Group 3, RAC.

13. Thomas G. Manning, *The Standard Oil Company: The Rise of a National Monopoly* (New York: Holt, Rinehart and Winston, 1960), 1. Although his brother Frank served in the Union Army, John instead paid for substitutes so he could continue his business. See Segall, *Rockefeller*, 28; Chernow, *Titan*, 69.

14. Early drillers applied imaginative names such as Oildorado to petroleum fields, and journalists invented similar names that recalled the gold and silver

booms in the US West. The title of Brian Black's recent study of the early oil economy borrows from this naming tradition. See Black, *Petrolia*, 65.

15. Black, *Petrolia*, 348, 351; J. T. Jenkins, *A History of the Whale Fisheries* (London: H. F. & G. Witherby, 1921), 232–35, 242; Peter Applebome, "They Used to Say Whale Oil Was Indispensible, Too," *New York Times*, August 3, 2008. For a recent environmental history of the American whaling industry, see Donald Worster, *Shrinking the Earth: The Rise and Decline of American Abundance* (New York: Oxford University Press, 2016), 57–73.

16. *Cleveland Morning Leader*, March 16, 1854, 3. See also Lance E. Davis, Robert E. Gallman, and Karin Gleiter, *In Pursuit of Leviathan: Technology, Institutions, Productivity, and Profits in American Whaling, 1816–1906* (Chicago: University of Chicago Press, 1997), 356.

17. Henry H. Townshend, *New Haven and the First Oil Well* (New Haven: Privately printed, 1934), 2. See also Ida M. Tarbell, *The History of the Standard Oil Company* (Gloucester: Peter Smith, 1963), 1:5.

18. J. T. Henry, *The Early and Later History of Petroleum* (New York: Burt Franklin, 1873), 1:53. See also Townshend, *New Haven*, 18; Paul Lucier, "The Professional and the Scientist in Nineteenth-Century America," *Isis* 100 (2009): 699–732.

19. Celia Campbell-Mohn, ed., *Environmental Law: From Resources to Recovery* (St. Paul: West Publishing, 1993), 700; *Westmoreland and Cambria Natural Gas Co. v. Dewitt*, 18 A. 724 (Pa. 1889). See also Harold F. Williamson and Arnold R. Daum, *The American Petroleum Industry: The Age of Illumination, 1859–1899* (Evanston: Northwestern University Press, 1959), 92–93, 101. On the informal use of the Rule of Capture, see Campbell-Mohn, *Environmental Law*, 700; David O. Whitten and Bessie E. Whitten, *The Birth of Big Business in the United States, 1860–1914: Commercial, Extractive, and Industrial Enterprise* (Westport, CT: Praeger, 2006), 138.

20. Townshend, *New Haven*, 29.

21. Hawke, "Interview," 43; "Mr. R comments on T.J. Gallagher," Inglis Notes, folder 12, box 4, Record Group 3, RAC.

22. "Mr. R comments on T.J. Gallagher," Inglis Notes, folder 12, box 4, Record Group 3, RAC. See also Allan Nevins, *John D. Rockefeller: The Heroic Age of American Enterprise* (New York: Charles Scribner's Sons, 1940), 1:178.

23. Hawke, "Interview," 87. See also William Donohue Ellis, *The Cuyahoga* (Dayton, OH: Landfall Press, 1975), 154. On the number of refineries, see Nevins, *John D.*, 1:178. On barrel and petroleum costs, see Allan Nevins, *Study in Power: John D. Rockefeller, Industrialist and Philanthropist* (New York: Charles Scribner's Sons, 1953), 1:33, 21. For Rockefeller's vertical integration of forests for barrels, see JDR Interview, September 3, 1917, Folder 3-8, box 4, Record Group 3, RAC.

24. Chernow, *Titan*, 78–79. See also Tarbell, *Standard Oil Company*, 1:39; Daniel Yergin, *The Prize: The Epic Quest for Oil, Money, and Power* (New York: Simon and Schuster, 1992), 35; Nevins, *Study in Power*, 1:37.

25. *Cleveland Leader*, April 24, 1877, 7, and April 21, 1877, 8; Nevins, *Study in Power*, 1: 33–34.

26. "Unknown 40-year veteran of SO, Nov. 4 1921," folder 11, box 4, Record Group 3, RAC. See also Lake Shore Crude Oil Transportation Company Records (MS 3494), folder 1, box 1, WRHS. For the cooperage report, see George H. Hopper to JDR, September 5, 1885, in folder 423, box 57, Record Group 1, RAC.

27. Edwin Cowles to Samuel Cowles, March 18, 1860, folder 3, box 1, Cowles-Hutchinson Papers, WRHS.

28. Edwin Cowles to Samuel Cowles, November 30, 1867, folder 4, box 1, Cowles-Hutchinson Papers, WRHS; Frank Rockefeller to JDR, January 21, 1888, in folder 473, box 64, Series C, RAC. On newspaper interests, see Chernow, *Titan*, 212.

29. O. H. Payne to JDR, December 31, 1884, folder 467, box 63, and F. B. Squire to JDR, March 21, 1887, folder 493, box 67, both in Series C, RAC. See also Chernow, *Titan*, 50, 240. For the political connection to Governor Foster, see Chas. W. Foster to JDR, July 4, 19, 1884, January 2, 1891, in folder 419, box 57, Record Group 1, RAC.

30. Chas. W. Foster to JDR, January 4, 1887, folder 419, box 57, Record Group 1, RAC.

31. Chernow, *Titan*, 132.

32. Chernow, *Titan*, 124–25, 184.

33. For his mother's admonishment, see "Interview with Mrs. John Wilcox, daughter of Jacob Rockefeller (1917)," on his boyhood memories of Lake Owasco, see JDR Interview, August 31, 1917, both in folder 3-8, box 4, Record Group 3, RAC. On driving horses, see "JDR Interview," October 7, 1917, folder 9, box 4. For Forest Hill business, see JDR to W. B. Smith, October 3, 1918, W. B. Smith to JDR, October 10, 1918, in folder 397, box 50, Series A, RAC.

34. Although his work focuses primarily on electrification, historian Wolfgang Schivelbush's history of artificial light parallels the cultural impact of kerosene illumination during the nineteenth century. See Wolfgang Schivelbush, *Disenchanted Night: The Industrialization of Light in the Nineteenth Century* (Berkeley: University of California Press, 1995).

35. Harold C. Livesay, "From Steeples to Smokestacks: The Birth of the Modern Corporation in Cleveland," in *The Birth of Modern Cleveland, 1865–1930*, ed. Thomas F. Campbell and Edward M. Miggins (Cleveland: Western Reserve Historical Society, 1988), 61.

TWO | FIRE

1. *Peter G. Golden vs. The Standard Oil Company, in Annals of Cleveland Court Record Series, 1875–1877* (Cleveland: US Works Progress Administration, Ohio, 1939), 10:222–23.

2. Stephen J. Pyne, *Fire: Nature and Culture* (London: Reaktion Books, 2012), 23, 159; Black, *Petrolia*, 81.

3. Williamson and Daum, *American Petroleum Industry*, 137, 168.

4. John T. Flynn, *God's Gold: The Story of Rockefeller and His Times* (New York: Harcourt, Brace, 1932), 120.

5. Townshend, *New Haven*, 24 (quotation), 29.

6. Chernow, *Titan*, 101; Flynn, *God's Gold*, 103–4.

7. William L. Leffler, *Petroleum Refining in Nontechnical Language*, 3rd ed. (Tulsa: PennWell, 2000), 12, 39, 49–51; George Armistead, Jr., *Safety in Petroleum Refining and Related Industries* (New York: John G. Simmonds, 1950), 251.

8. Williamson and Daum, *American Petroleum Industry*, 193. Following the Bennehoff Run fire and other conflagrations of the 1860s, producers also added space between storage tanks (350–400 feet instead of 200 or less), see Hidy and Hidy, *Pioneering*, 82.

9. Hidy and Hidy, *Pioneering*, 82; Lloyd, *Wealth Against Commonwealth*, 162.

10. Black, *Petrolia*, 81; W. P. Thompson to JDR, April 9, June 30, 1885, folder 511, box 69, Series C, RAC.

11. Christopher F. Jones, *Routes of Power: Energy and Modern America* (Cambridge, MA: Harvard University Press, 2014), 120–21. See also Albert Z. Carr, *John D. Rockefeller's Secret Weapon* (New York: McGraw-Hill, 1962), 32.

12. Williamson and Daum, *American Petroleum Industry*, 165.

13. Lloyd, *Wealth Against Commonwealth*, 139. See also Williamson and Daum, *American Petroleum Industry*, 168.

14. *Cleveland Plain Dealer*, July 23, 1874, 3.

15. On rebates and drawbacks, see Chernow, *Titan*, 113; H. W. Brands, *American Colossus: The Triumph of Capitalism, 1865–1900* (New York: Doubleday, 2010), 84–85; Yergin, *The Prize*, 39; Nevins, *Study in Power*, 1:107.

16. Frank R. Rockefeller to JDR, December 6, 1887, folder 257, box 34, Series C, RAC.

17. R. J. Forbes, *More Studies in Early Petroleum History, 1860–1880* (Leiden: E. J. Brill, 1959), 117.

18. Tarbell, *Standard Oil Company*, 1:19; James G. Speight, *The Chemistry and Technology of Petroleum*, 3rd ed. (New York: Marcel Dekker, 1999), 7; "Annual Message of the Mayor of Cleveland, Hon. Stephen Buhrer, to the City Council, April 11th, 1871," in Cleveland, Ohio, *Reports of the Departments of the Govern-*

ment of the City of Cleveland for the Year Ending December 31, 1871 (Cleveland: Leader Book and Job Office, 1872), 8. These annual reports will henceforth be cited as Cleveland, Ohio, *Reports*, with the relevant date.

19. JDR interview, September 20, 1917, Inglis Notes, folder 9, box 4, Record Group 3, RAC. See also Chernow, *Titan*, 101; for the fires of 1873, see *Cleveland Leader*, November 24, December 1, 15, 25, 1873, all page 4.

20. Cleveland, Ohio, *Reports, 1865* (Cleveland: Leader Company, 1866), 7–8, also 69, 85, 83. See also Cleveland, Ohio, *Reports, 1879* (Cleveland: Wiseman and Harvey Printers, 1880), 505; David D. Van Tassel and John J. Grabowski, eds., *The Encyclopedia of Cleveland History* (Bloomington: Indiana University Press, 1987), 232–33; William Ganson Rose, *Cleveland: The Making of a City* (Cleveland: World Publishing, 1950), 439; Robert Chester, "Manufacturing Danger: The Perils of Place" (unpublished dissertation, University of California, Davis, 2009), 22, in author's possession.

21. "Directions and Instructions for Fire Alarm Telegraph," in Cleveland, Ohio, *Reports, 1872* (Cleveland: Waechter am Erie Printing Co., 1872), 231.

22. *Cleveland Leader*, February 24, 1880, 8.

23. W. P. Thompson to JDR, March 1, 1883 ("by reducing" and "satisfactory rates"), December 20, 1884 ("in the event"), folder 509, box 69, Series C, RAC.

24. *Cleveland Leader*, February 5, 1883, 1, 6.

25. *Cleveland Leader*, February 5, 1883, 6.

26. *Cleveland Leader*, February 5, 1883, 6.

27. *Cleveland Leader*, February 5, 1883, 8.

28. Williamson and Daum, *American Petroleum Industry*, 290, 471; Rose, *Cleveland*, 362.

29. Alice Taylor, *Quench the Lamp* (New York: St. Martin's Press, 1990), 110–11.

30. Forbes, *More Studies*, 108–9.

31. Forbes, *More Studies*, 130–31; Nevins, *Study in Power*, 2:45, 47.

32. Hidy and Hidy, *Pioneering*, 117.

33. H. A. Hutchins to Col. O. H. Payne, April 10, April 7, 1879, folder 426, box 57, subseries B, Series C, Record Group 1, RAC.

34. Geo. F. Gregory to JDR, New York, February 16, 1884, folder 421, box 57, and W. P. Thompson to JDR, January 26, 1885, folder 510, box 69, both in subseries B, Series C, Record Group 1, RAC.

35. F. Q. Barstow to The Executive Committee (SOT), July 27, 1888, F. Q. Barstow to David S. Cowles (sec. Domestic Trade Committee), August 2, 1888; sales report, H. C. Folger, Jr., to H. H. Rogers, August 3, 1888, all in folder 417, box 56, Record Group 1, RAC.

36. F. B. Squire to JDR (most likely), June 11, 1889, folder 495, box 67, JDR to F Q. Barstow, November 30, 1883, folder 482, box 65, O. H. Payne to JDR, November 18, 1884, folder 467, box 63, all in Series C, RAC.

37. Frank Rockefeller to JDR, June 12, 1889, folder 478, and February 25, 1892, folder 481, both in box 65, Series C, RAC.

38. RAF to ARF (probably Frank Rockefeller to Sam Andrews), internal telegram, April 15, 1868, in book 394, subseries 19, Series L, RAC.

39. JDR Interview with Inglis, May 15, 1919, folder 10, box 4, Record Group 3, RAC. See also H. C. Folger, Jr., to A. M. McGregor, May or June 1890, memo 1360, folder 416, box 56, Record Group 1, RAC.

40. J. G. W. Cowles to George D. Rogers, July 24, 1888, folder 67, box 9, Record Group 1, RAC. See also Hidy and Hidy, *Pioneering*, 297–99.

41. Margaret Hindle Hazen and Robert M. Hazen, *Keepers of the Flame: The Role of Fire in American Culture, 1775–1925* (Princeton, NJ: Princeton University Press, 1992), 74; Steinberg, *Down to Earth*, 65–66.

42. Cleveland, Ohio, *Reports, 1878* (Cleveland: Wiseman and Harvey, 1879), 471; "Paying the Piper," *Cleveland Leader*, July 17, 1874, 4; Stephen J. Pyne, *Fire: A Brief History* (Seattle: University of Washington Press, 2001), 114. Charles Rosenberg's history of cholera documents the evolution in understanding of the disease in nineteenth-century America, which mirrors the development of a rational understanding of fire. Cholera was initially believed to originate in places of low morals, but by the 1860s observers had abandoned this view and adopted sanitary reforms that attacked the disease at its true source of transmission. See Charles E. Rosenberg, *The Cholera Years: The United States in 1832, 1849, and 1866* (Chicago: University of Chicago Press, 1962).

43. Cleveland, Ohio, *Reports, 1865*, 89; Cleveland, Ohio, *Reports, 1874* (Cleveland: Co-operative Printing Company, 1875), 353. Often, if a human actor was involved in the cause of the fire, regardless of its source, the fire department recorded such blazes as "accidents" or the result of "carelessness." Thus, a fire at the Standard Oil Works on October 30, 1874, which "did little damage[,] . . . was caused by carelessness." See *Cleveland Leader*, October 31, 1874, 7. Children playing with matches caused two fires in 1865 and ten in 1874.

44. *Cleveland Leader*, June 27, 1874, 8; *Cleveland Leader*, July 29, 1872, 4.

45. Captain J. H. Thomson and Boverton Redwood, *Handbook on Petroleum* (London: Charles Griffin, 1901), 175. See also Lloyd, *Wealth Against Commonwealth*, 409.

46. For the federal fire test standard and Dr. Chandler's study, see Stephen Farnum Peckham, *Report on the Production, Technology, and Uses of Petroleum and Its Products* (Washington, DC: Government Printing Office, Department of the

Interior, 1885), 236–37; Hidy and Hidy, *Pioneering*, 99; Cleveland, Ohio, *Reports, 1873* (Cleveland: Fairbanks, Benedict, 1874), 236.

47. "The 'Non-explosive' Committee," *Cleveland Leader*, January 9, 1872, 1, 4; "Legislating on Coal Oils," *Cleveland Leader*, February 1, 1872, 2.

48. *Cleveland Leader*, January 7, 1875, 8.

49. *Cleveland Leader*, February 26, 1876, 7; "Fire Commissioner's Report," in Cleveland, Ohio, *Reports, 1876* (Cleveland: A. W. Fairbanks and Company, 1877), 160. Exploding lamps were responsible for 22 of 224 fires of known causes. See Cleveland, Ohio, *Reports, 1876*, 159; Lloyd, *Wealth Against Commonwealth*, 410.

50. *Cleveland Leader*, May 12, 1876, 7, June 10, 1876, 7, December 15, 1873, 4.

51. Forbes, *More Studies*, 139.

52. "Seventh Annual Report of the Chief Engineer of the Cleveland Steam Fire Department, for the Fiscal Year, Ending March 31st, 1871," in Cleveland, Ohio, *Reports, 1871*, 85; "City Auditor's Report," in Cleveland, Ohio, *Reports, 1874*, 9–10. A year later the city's debt would reach $5.1 million. See Cleveland, Ohio, *Reports, 1875* (Cleveland: Co-operative Printing Company, 1876), xiv.

53. "Inaugural Address of Hon. N. P. Payne, Mayor of Cleveland," Cleveland, Ohio, *Reports, 1874*, xxv.

54. "Inaugural Address of Hon. R. R. Herrick, Mayor," Cleveland, Ohio, *Reports, 1878*, 56–57.

55. "Mayor's Annual Message, R. R. Herrick," Cleveland, Ohio, *Reports, 1880* (Cleveland: Home Companion Publishing Company, 1881), 13. See also N. S. Cobleigh, *The Manufactures, Trade and Commerce of Cleveland, 1880–1881* (Cleveland: Short and Forman, 1881), 21.

56. Lloyd, *Wealth Against Commonwealth*, 414. See also Hidy and Hidy, *Pioneering*, 592.

57. *Michigan Law Review* 3, no. 6 (April 1905): 494.

58. Peckham, *Report*, 237.

59. *Peter G. Golden vs. The Standard Oil Company*, in the *Annals of Cleveland Court Record Series, 1875–1877* (Cleveland: 1939), 10:223.

60. JDR, *Random Reminiscences*, 58. See also Nevins, 2:106–7, 111. See J. R. McNeill, *Something New under the Sun: An Environmental History of the Twentieth-Century World* (New York: Norton, 2000), 297–98, for the emergence of the "Motown Cluster," or the oil-driven mode of production that marked the twentieth-century human relationship with nature.

61. Cleveland, Ohio, *Reports, 1868* (Cleveland: Leader Book and Job Office, 1869), 67; Cleveland, Ohio, *Reports, 1896* (Cleveland: Cleveland Printing and Publishing, 1897), 806–10; Yergin, *The Prize*, 50.

THREE | WATER

1. "Through the Tunnel," *Cleveland Leader*, February 23, 1874, 8.

2. For an early social history of disease, see Rosenberg, *Cholera Years*. For a history of New York City's efforts to tap its hinterland watershed to serve a burgeoning urban population, see Gerard T. Koeppel, *Water for Gotham: A History* (Princeton, NJ: Princeton University Press, 2001). See also Daniel D. Jackson, *Report on the Sanitary Condition of the Cleveland Water Supply on the Probable Effect of the Proposed Changes in Sewage Disposal and on the Various Sources of Typhoid Fever in Cleveland* (Cleveland: City of Cleveland, 1912), 13.

3. Abraham Gesner, *A Practical Treatise on Coal, Petroleum and Other Distilled Oils*, 2nd ed. (New York: Augustus M. Kelley, 1968), 158.

4. JDR interview, September 4, 1917, folder 9, box 4, Record Group 3, RAC. See also Maxcine J. Japour, *Petroleum Refining and Manufacturing Processes* (Los Angeles: Wetzel Publishing, 1939), 54; Nevins, *Study in Power*, 136; Grasselli Acid Works Account Book, December 3, 1868, through February 27, 1869, folder 68, box 4, Grasselli Family Papers, WRHS.

5. Virginia Elwood-Akers, *Caroline Severance* (New York: iUniverse Inc., 2010), 29. See also J. G. W. Cowles to JDR, December 16, 1887, folder 66, box 9, Record Group 1, RAC. For more on Dr. Seelye's water cure see Elwood-Akers, *Caroline Severance*, 9; Alan Dutka, *Cleveland Calamities: A History of Storm, Fire, and Pestilence* (Charleston, SC: History Press, 2014), 115–18.

6. "Our Beautiful Drinking Water," *Cleveland Leader*, March 6, 1866, 2. See also "The Water We Drink," *Cleveland Leader*, December 20, 1866, 4.

7. "Chicago and Her Tunnel," *Cleveland Leader*, December 3, 1866, 2.

8. "Special Report of the Superintendent and Engineer upon the Quality of the Water of Lake Erie at the Present Time," in Cleveland, Ohio, *Reports, 1866* (Cleveland: Fairbanks, Benedict and Co., Printers, Herald Office, 1867), 131, 131, 132. See also "The Claims of Our Vessel Owners vs. the Oil Refineries," *Cleveland Leader*, November 7, 1867, 2.

9. "The Oil Nuisance," *Cleveland Leader*, February 13, 1868, 4.

10. Morton J. Horwitz, *The Transformation of American Law, 1780–1860* (Cambridge, MA: Harvard University Press, 1977), 55 (quoting *Jackson v. Brownson* court comment).

11. Horwitz, *Transformation, 1780–1860*, 41 (quoting Lemuel Shaw, Chief Justice of the Massachusetts Supreme Court in *Cary v. Daniels*), 37. For an environmental history centered on the relationships between the law, nature, and private capital in New England, see Steinberg, *Nature Incorporated*.

12. Horwitz, *Transformation, 1780–1860*, 99; James Willard Hurst, *Law and the Conditions of Freedom in the Nineteenth-Century United States* (Madison: Univer-

sity of Wisconsin Press, 1956), 6. Arthur F. McEvoy's pioneering study of California fisheries provides an aquatic example of how changes in the law contributed to the "tendency of market forces to sunder the ecological bonds between natural resources and their environments." Arthur F. McEvoy, *The Fisherman's Problem: Ecology and Law in the California Fisheries, 1850–1980* (New York: Cambridge University Press, 1986), 119.

13. "Our Filthy Drinking Water," *Cleveland Leader*, December 23, 1867, 2. See also "Confidential Report by the Committee on Industrial Development," November 14, 1924, The Greater Cleveland Growth Association Records (MS 3471), folder 159, WRHS; "The Oil Question Again," *Cleveland Leader*, November 7, 1867, 1; "The Oil Nuisance," *Cleveland Leader*, November 14, 1867, 2. The historian Edward Tenner describes a Revenge Effect as arising when a new technology "induces behavior which appears to cancel out the very reason for using it." Edward Tenner, *Why Things Bite Back: Technology and the Revenge of Unintended Consequences* (New York: Vintage Books, 1997), 7.

14. "Our Filthy Drinking Water," *Cleveland Leader*, December 23, 1867, 2.

15. "Song of the Sick Water-Nymph," *Cleveland Leader*, February 13, 1868, 4.

16. Cleveland, Ohio, *Reports, 1873*, 269.

17. Cleveland, Ohio, *Reports, 1868*, 125; "The Oil Nuisance," *Cleveland Leader*, September 2, 1868, 4. See also Cleveland, Ohio, *Reports, 1873*, 269.

18. "The Drinking Water," *Cleveland Leader*, February 16, 1872, 1.

19. Cleveland, Ohio, *Reports, 1874*, ix. See also Cleveland, Ohio, *Reports, 1875*, xiv.

20. For a comprehensive history of this moment, see M. John Lubetkin, *Jay Cooke's Gamble: The Northern Pacific Railroad, the Sioux, and the Panic of 1873* (Norman: University of Oklahoma Press, 2006); *Cleveland Leader*, April 26, 1877, 4; Winston Chrislock, "Cleveland's Czechs," *Identity, Conflict, and Cooperation: Central Europeans in Cleveland, 1850–1930*, ed. David C. Hammack, Diane L. Grabowski, and John J. Grabowski (Cleveland: Western Reserve Historical Society, 2002), 101.

21. "The Workingmen," *Cleveland Leader*, April 25, 1877, 4, 8; *Cleveland Leader*, May 14, 1877, 8.

22. *Cleveland Leader*, February 24, 1875, 7. See also *Cleveland Leader*, February 19, 1875, 8.

23. *Cleveland Leader*, July 28, 1875, 7, August 18, 1875, 7, also August 25, 1875, 8.

24. *Cleveland Leader*, July 30, 1877, 4; "Inaugural Address of Hon. R. R. Herrick, Mayor," in Cleveland, Ohio, *Reports, 1878*, 11–12, 57.

25. "Anonymous," letter to the editor, *Cleveland Leader*, May 14, 1875, 7; "Report of the Health Department," and "Mayor's Annual Message, R. R. Herrick," Cleveland, Ohio, *Reports, 1879*, 428, 44.

26. "Mayor's Annual Message, R. R. Herrick," and W. B. Rezner, M.D., "Annual Report of the Health Department," in Cleveland, Ohio, *Reports, 1880*, 12–13, 453–56. See also "Awful Offal," *Cleveland Leader*, February 24, 1880, 8.

27. "Civil Engineer's Report, B. F. Morse," Cleveland, Ohio, *Reports, 1881* (Cleveland: Home Companion Publishing, 1882), 201–2.

28. "Report of the Health Department," Cleveland, Ohio, *Reports, 1890* (Cleveland: Cleveland Printing and Publishing, 1891), 570. The health department report during the 1886 term acknowledged the "inexhaustible" nature of the lake water but pointed to the "increasing amount of excrementitious material" entering all the lakes from various rivers and port cities, challenging the planners to consider the possibility of confederated action. "Report of the Health Department," Cleveland, Ohio, *Reports, 1886* (Cleveland: Peerless Printing, 1887), 780–81. A year later the health department hired professor A. W. Smith of the Case School of Applied Science to conduct a thorough study of the lake water quality. The dangers outlined in the report were soft-pedaled by the department, which declared "the public water supply . . . [is] of excellent quality," downplaying the "rare" "exceptions." Cleveland, Ohio, *Reports, 1887* (Cleveland: The Cleveland Printing and Publishing, 1888), 9–10. The River and Sewer Commission, according to the civil engineer's report of 1887, gave "earnest attention" to the "question of the pollution of the Cuyahoga river," which promised "a satisfactory solution of this great problem may be confidently expected within a reasonable measure of time." "Report of City Civil Engineer," Cleveland, Ohio, *Reports, 1887*, 391.

29. Cleveland, Ohio, *Reports, 1892* (Cleveland: Cleveland Printing and Publishing, 1893), 774–75; "Chas. J. Wheeler, Food Inspector," Cleveland, Ohio, *Reports, 1896*, 778.

30. "Report on the Proposed Extension of Water Works Tunnel, Intercepting Sewerage System and River Flushing Tunnel, presented by Rudolph Hering, George H. Benzenberg, and Desmond Fitzgerald, Expert Engineers to the Mayor and Council of the City of Cleveland, February 4, 1896," Cleveland, Ohio, *Reports, 1895* (Cleveland: Brooks, 1896), lx, lxvi, lxvi, lxvii. See also M. E. Rawson, Chief Engineer, in Cleveland, Ohio, *Reports, 1897* (Cleveland: Brooks, 1898), 504–5.

31. *Cleveland Leader*, January 24, 1868, 4. Standard's ledger for the Bayonne refinery in 1902 show that outlays for acid cost the company $169,486.77 while labor costs over the same period amounted to $139,953.95. See "Analysis of Business of Standard Oil Co. of New Jersey, Bayonne for 12 months ending December 31, 1902," folder entitled Bayonne and Bayway, 1899–1911, in box 2.207/K100C, ExxonMobil Archives, Briscoe Center for American History, University of Texas at Austin.

32. Jamie Benidickson, *The Culture of Flushing* (Vancouver: UBC Press, 2007), 7; *Cleveland Leader*, July 22, 1885, 5.

33. Cleveland, Ohio, *Reports, 1892*, 165. It should be remembered that the daily consumption figures represent an average of the yearly total. Actual figures fluctuated with weather and seasonal use. For instance, summer months saw far higher daily rates than these figures suggest. Public works officers recorded daily consumption rates as high as 48 million gallons during the summer months of 1890, a full two years before the average daily consumption reached 32.2 million gallons. See Cleveland, Ohio, *Reports, 1891* (Cleveland: The Cleveland Printing and Publishing Co., 1892), 42.

34. Willis E. Sibley, "Water System," in Van Tassel and Grabowski, *Encyclopedia*, 1029–32.

35. Inglis interview cited in Hawke, "Interview," 905. See also Ella Grant Wilson, *Famous Old Euclid Avenue of Cleveland* (Cleveland: privately printed, 1937), 2:146.

36. "The Business-Men's Meeting to Be Held Today," *Cleveland Leader*, August 9, 1899, 10; "Against the Boycott and Mob Violence," *Cleveland Leader*, August 10, 1899, 1. For brief biographies of John Guiteau Welch Cowles, see *The Book of Clevelanders: A Biographical Dictionary of Living Men of the City of Cleveland* (Cleveland: Burrows Bros., 1914), 64; William C. Barrow, *The Euclid Heights Allotment* (Masters thesis, Cornell University, Ithaca, New York, 1996); Joseph P. Smith, ed., *History of the Republican Party* (Chicago: The Lewis Publishing Co., 1898).

37. Samuel Orth, *A History of Cleveland*, vol. 1 (Chicago: S. J. Clarke, 1910); "Explosion in Tunnel Kills Five," *Los Angeles Herald*, August 22, 1901.

38. R. Winthrop Pratt, "A New Move for Water Filtration at Cleveland, Ohio," *Engineering News* 69, no. 21 (May 22, 1913): 1078. See also George C. Whipple, *Report on the Quality of the Water Supply of the City of Cleveland, Ohio* (Cleveland: Division of Water Works, 1905), 7–11.

39. Thomas P. Hughes, "Technological Momentum," in *Does Technology Drive History? The Dilemma of Technological Determinism*, ed. Leo Marx and Merritt Roe Smith (Cambridge, MA: MIT Press, 1994), 101–13; Jones, *Routes of Power*. See also R. Winthrop Pratt, "Sewage Disposal Investigations at Cleveland," *Engineering News* 69, no. 7 (February 13, 1913): 287–94; *Ohio Public Health Journal* 5, no. 6 (June 1915): 784; Richard H. Thaler, "Toward a Positive Theory of Consumer Choice," *Journal of Economic Behavior and Organization* 1, no. 1 (1980): 39–60.

40. Cobleigh, *Manufactures*, 111. Statistics are from the Greater Cleveland Growth Association Records (MS 3471), folder 68, WRHS.

41. Martin V. Melosi, *The Sanitary City* (Baltimore: Johns Hopkins University Press, 2000), 10; Samuel Taylor Coleridge, *The Rime of the Ancient Mariner* (New York: Harper and Bros., 1877). See also Morton J. Horwitz, *The Transformation of American Law, 1870–1960: The Crisis of Legal Orthodoxy* (New York: Oxford University Press, 1992), 4, 66.

FOUR | AIR

1. "Report of Smoke Inspector, James McLaren," *Annual Reports, City of Cleveland*, 1898 (Cleveland: The Cleveland Ptg. & Pub. Co., 1899), 806.

2. Historian Sam Warner's *Streetcar Suburbs* remains the authoritative history of the development of the early suburb, which Warner attributes to the development of electricity and the reduced cost of affiliated materials. His thesis pertains to Boston but fails to account for the "push factors" of deteriorating environmental health that led many affluent Clevelanders to find solace in bucolic suburban real estate. See especially Sam B. Warner Jr., *Streetcar Suburbs: The Process of Growth in Boston, 1870–1900* (Cambridge, MA: Harvard University Press, 1962), 28; Kenneth T. Jackson, *Crabgrass Frontier: The Suburbanization of the United States* (New York: Oxford University Press, 1985), 285.

3. "Hon. Geo. B. Senter, Annual Message," in Cleveland, Ohio, *Reports, 1864* (Cleveland: Fairbanks, Benedict and Co., City Printers, Herald Office, 1865), 6; Rend quoted in McNeil, *Something New*, 59.

4. The classic text on the culture of Western conquest is Henry Nash Smith, *Virgin Land: The American West as Symbol and Myth* (Cambridge, MA: Harvard University Press, 1950). For an introduction to the New Western history, see Limerick, *Legacy of Conquest*; Clyde A. Milner, Patricia Nelson Limerick, and Charles E. Rankin, eds., *Trails: Toward a New Western History* (Lawrence: University of Kansas Press, 1991); and Richard Slotkin's trio of books on the West as myth, but especially *The Fatal Environment: The Myth of the Frontier in the Age of Industrialization, 1800–1890* (Norman: University of Oklahoma Press, 1998). For the viaduct, see W. Scott Robison, ed., *History of the City of Cleveland: Its Settlement, Rise and Progress* (Cleveland: Robison and Cockett, 1887), 276. Robison records that Mayor W. G. Rose reported the viaduct was valued at $2,135,000.00 in 1878.

5. A weather inversion causing the air pollution from the American Steel and Wire Company in Donora, Pennsylvania, to accumulate for a period of four days during October 1948 killed twenty people and caused severe respiratory tract illness among 42 percent of the city's population. The incident led the state of Pennsylvania to call on the US Public Health Service to conduct a thorough investigation within two months of the disaster. This was the first investigation of its kind in the United States. An army of public health nurses canvassed the population door-to-door; the local hospital's health records were reviewed; blood, teeth, and urine samples were taken from residents and animals in the area; and autopsies were performed on the bodies of several of the incident's victims. The American Steel and Wire Company defended itself in court by claiming the disaster was "an act of God." McNeil, *Something New*, 70. The event and the subsequent study provided scientists and public health officials with the most thorough data to date

concerning the adverse health effects of air pollution. See Harry Heimann, "Effects of Air Pollution on Human Health," World Health Organization, *Air Pollution* (New York: Columbia University Press, 1961), 159–220.

6. See Moyer D. Thomas, "Effects of Air Pollution on Plants," World Health Organization, *Air Pollution* (New York: Columbia University Press, 1961), 234; Heimann, "Effects of Air Pollution," 187.

7. Heimann, "Effects of Air Pollution," 194.

8. Constance F. Woolson, "Round by Propeller," *Harper's New Monthly Magazine* 45, no. 268 (September 1872): 520, 521.

9. Constance F. Woolson, "Round by Propeller," 522, 524.

10. *Cleveland Leader*, June 28, 1870, 2.

11. *Cleveland Leader*, July 31, 1880, 4.

12. *Cleveland Leader*, June 30, 1870, 2.

13. "The Slough of Despond," *Cleveland Leader*, November 14, 1874, 7; *Cleveland Leader*, August 18, 1875, 7. Somewhat ironically, the pro-business *Leader* demanded at the time that the "action be spiked at once" lest business flee the city. *Cleveland Leader*, June 30, 1870, 2. See also Chapman, *Cleveland*, 116; *Cleveland Leader*, August 25, 1875, 8.

14. *Cleveland Leader*, August 2, 1880, 5. See also Philip W. Porter, *Cleveland: Confused City on a Seesaw* (Columbus: Ohio State University Press, 1976), 3.

15. "Letter from J.C. Arthur, Prof. of Botany, Purdue Univ, to Sec. of Agriculture," in "Report issued by Assistant Secretary of Agriculture, Edwin Willits, 25 September 1891," Cleveland, Ohio, *Reports, 1891*, 55, 61, 62.

16. "Letter from J.C. Arthur," in Cleveland, Ohio, *Reports, 1891*, 59.

17. "Letter from B. E. Fernow, Chief of Forestry Division, to Secretary of Agriculture, 18 September 1891," Cleveland, Ohio, *Reports, 1891*, 56; "Letter from J.C. Arthur," Cleveland, Ohio, *Reports, 1891*, 59–60; "Letter from B. E. Fernow," Cleveland, Ohio, *Reports, 1891*, 57–58.

18. "Report of the Smoke Inspector," Cleveland, Ohio, *Reports, 1884* (Cleveland: Peerless Printing Company, 1885), 457; "Mayor's Annual Message to the City Council," Cleveland, Ohio, *Reports, 1884*, xx.

19. "Report of the Smoke Inspector," Cleveland, Ohio, *Reports, 1884*, 458; "Department of Police, Health Division, Dr. Geo. F. Leick, Health Officer," Cleveland, Ohio, *Reports, 1893* (Cleveland: J. B. Savage, Print, 1894), 663–64.

20. "Department of Police, Health Division, Dr. Geo. F. Leick, Health Officer," in Cleveland, Ohio, *Reports, 1893*, 663–64; Ralph E. Chapman and George G. Cummings, "A Study of the 'Sootfall' in the City of Cleveland, Ohio" (BS thesis, Case School of Applied Science, Cleveland, 1924), 23–24; "St. Alexis Hospital Medical Center," and Willis E. Sibley, "Water System," both in Van Tassel and Grabowski, *Encyclopedia*, 85; Angela Gugliotta, "'Hell with the Lid Taken Off': A

Cultural History of Air Pollution—Pittsburgh" (unpublished dissertation, University of Notre Dame, 2004); L.A. Richards, ed., *Diagnosis and Improvement of Saline and Alkali Soils* (Washington, DC: US Department of Agriculture, 1954), 158.

21. *Cleveland Leader*, July 31, 1880, 5.

22. "Report of Committee on Walworth Run Nuisance," Cleveland, Ohio, *Reports, 1873*, 269–70.

23. "Report of Committee on Walworth Run Nuisance," Cleveland, Ohio, *Reports, 1873*, 20. See also Cleveland, Ohio, *Reports, 1897*, 499.

24. Cleveland, Ohio, *Reports, 1875*, 698; Cleveland, Ohio, *Reports, 1878*, 388.

25. *Cleveland Leader*, February 24, 1880, 8.

26. *Cleveland Leader*, July 30, 1880, 6.

27. "Report of the Health Department," Cleveland, Ohio, *Reports, 1881*, 421; "That Stink," *Cleveland Leader*, August 2, 1880, 5. For Andrews, see Wilson, *Euclid Avenue*, 2:1–2.

28. Interview with anonymous employee, November 10, 1921, Inglis Notes, folder 11, box 4, Record Group 3, RAC.

29. JDR, *Random Reminiscences*, 26.

30. Joseph W. Ernst, ed., *"Dear Father"/"Dear Son": Correspondence of John D. Rockefeller and John D. Rockefeller, Jr.* (New York: Fordham University Press, 1994), 19; FR to JDR, August 16, 1888, folder 257, box 34, Series C, RAC. See also Cleveland, Ohio, *Reports, 1891*, 58.

31. JDR to Cleveland City Council, October 29, 1897, Letterbooks, vol. 46, page 78, Record Group 1, John D. Rockefeller Papers (JDR Papers), Rockefeller Family Archives, RAC. See also Rose, *Cleveland*, 573.

32. *The Cleveland Chamber of Commerce, Reports and Proceedings: 46th Annual Meeting, April 17th, 1894*, The Greater Cleveland Growth Association Records (MS 3471), cont. 68, WRHS.

33. *Cleveland Recorder*, November 9, 1897, 29. The remainder of newspaper citations in this chapter refer to the page number associated to the 135-page album compiled by the Cleveland Park Board Reorganization Association in 1898 and held at the Cleveland Public Library.

34. *Cleveland Leader*, December 29, 1897, 43.

35. *Cleveland World*, February 10, 15, 1898, 6; James Harrison Kennedy, *A History of the City of Cleveland: Its Settlement, Rise, and Progress, 1796–1896* (Cleveland: The Imperial Press, 1896), 513.

36. *Cleveland Recorder*, November 9, 1897, 29; Hanna quoted in William Safire, *Safire's Political Dictionary* (New York: Oxford University Press, 2008), 237.

37. *Cleveland Recorder*, August 16, 1897, 18; *Cleveland Press*, September 16, 1897, 19. See also *Cleveland Press*, September 20, 1897, 19, 21.

38. *Cleveland Press*, January 6, 1898, 57.

39. *Cleveland Recorder*, September 28, 1897, 23; *Cleveland World*, December 22, 1897, 38.

40. *Cleveland Leader*, February 1, 1897, 30–31. See also *Cleveland Press*, February 18, 1898, 10.

41. *Cleveland Leader*, February 14, 1898, 3.

42. *Cleveland Leader*, December 12, 1897, 38.

43. *Cleveland World*, February 15, 1898, 5; *Cleveland Recorder*, February 15, 1898, 3; *Cleveland Press*, December 27, 1897, 41; *Cleveland Leader*, December 29, 1897, 43.

44. *Cleveland Leader*, February 6, 1898, 96.

45. *Cleveland Recorder*, February 15, 1898, 9. For a biography of James H. Hoyt, see George Irving Reed, ed., *Bench and Bar of Ohio: A Compendium of History and Biography*, vol. 2 (Chicago: Century Publishing and Engraving Company, 1897), 206–8; *Cleveland Leader*, February 6, 1898, 96, and February 13, 1898, 126.

46. *Cleveland Recorder*, February 14, 1898; *William Macey vs. The Standard Oil Company*, August 5, 1872, in *Annals of Cleveland Court Record Series*, vol. 8, *1871–1872* (Cleveland, 1939), 263. See also *Cleveland Leader*, July 31, 1884, 5.

47. *Cleveland Leader*, January 20, 1898, 64. See also Carol Poh Miller, *Cleveland Metroparks, Past and Present* (Cleveland: Cleveland Metroparks, 1992). Peter Witt, labeled an anarchist by opponents, attacked even Johnson insiders who failed to practice the reform they preached. See Hoyt Landon Warner, *Progressivism in Ohio, 1897–1917* (Columbus: Ohio State University Press, 1964), 65–66; also Carol Poh Miller, "Parks," in Van Tassel and Grabowski, *Encyclopedia*, 752–54. Several environmental histories of the early conservation movement focus attention on the class dimensions of Progressive reforms; see Robert Gottlieb, *Forcing the Spring: The Transformation of the American Environmental Movement* (Washington, DC: Island Press, 1993); Louis Warren, *The Hunter's Game: Poachers and Conservationists in Twentieth-Century America* (New Haven: Yale University Press, 1999); Karl Jacoby, *Crimes against Nature: Squatters, Poachers, Thieves, and the Hidden History of the American Conservation Movement* (Berkeley: University of California Press, 2003).

48. Hidy and Hidy, *Pioneering*, 434.

49. Cleveland, Ohio, *Reports, 1896*, 749; Cleveland, Ohio, *Reports, 1898* (Cleveland: Cleveland Printing. and Publishing, 1899), 806.

50. R. Dale Grinder, "The Battle for Clean Air: The Smoke Problem in Post–Civil War America," *Pollution and Reform in American Cities, 1870–1930*, ed. Martin V. Melosi (Austin: University of Texas Press, 1980), 98.

51. Cleveland, Ohio, *Reports, 1902* (Cleveland: A. S. Gilman, 1903), 941.

52. *Cleveland Leader*, March 2, 1904, 10.

FIVE | EFFICIENT EARTH

1. Alfred D. Chandler Jr., "The Standard Oil Company—Combination, Consolidation, and Integration," in *The Coming of Managerial Capitalism*, ed. Alfred D. Chandler Jr. (Homewood, IL: Richard D. Irwin, 1985), 349, 354, 360. For Cleveland's manufacturing limitations, see Nevins, *John D.*, 1:182–83; Nevins, *Study in Power*, 1:49. For figures on peak production see *Report of the Commissioner of Corporations on the Petroleum Industry, Part II: Prices and Profits* (Washington, DC: Government Printing Office, 1907), 103.

2. Chandler, "Standard Oil," 349, 354, 360. Three years after its discovery, Lima crude accounted for a staggering 35 percent of American crude production. Standard Oil, able to acquire producing lands and storage early, sought to break the power of independent Pennsylvania producers by using Lima crude as much as possible. For instance, its Bayonne refinery in New Jersey had exclusively refined Pennsylvania crude prior to 1885. By 1898, however, 7.1 million barrels of its annual 8 million barrel throughput originated from the Lima field. See Gerald T. White, *Formative Years in the Far West: A History of Standard Oil Company of California and Predecessors through 1919* (New York: Appleton-Century-Crofts, 1962), 195; Hidy and Hidy, *Pioneering*, 286; Williamson and Daum, *American Petroleum Industry*, 606. For the Pennsylvania producers' association, see Gilbert Holland Montague, "The Later History of the Standard Oil Company," *Quarterly Journal of Economics* 17, no. 2 (February 1903): 302–3.

3. Paul H. Giddens, *Standard Oil Company (Indiana): Oil Pioneer of the Middle West* (New York: Appleton-Century-Crofts, 1955), 10.

4. "Greatest in the World," *Chicago Tribune*, September 22, 1889, 9; U. G. Swartz, "Some Early Days of Whiting Refinery," *Stanolind Record* 4(9): 12.

5. Powell A. Moore, *The Calumet Region: Indiana's Last Frontier* (Indianapolis: Indiana Historical Bureau, 1959), 186; William Cronon, *Changes in the Land: Indians, Colonists, and the Ecology of New England* (New York: Hill and Wang, 1983), 230–35; T. H. Ball, *Northwestern Indiana: From 1800 to 1900* (Chicago: Donohue and Henneberry, 1900), 410–11; Moore, *Calumet Region*, 101–2; Giddens, *Standard Oil*, 12–13.

6. The location was named for a reckless railroad engineer, "Pap" Whiting, who had derailed his train taking a turn on his way to Chicago. See "How Chicago's Suburbs Were Planned and Named," *Chicago Tribune*, March 4, 1900, 46; also Moore, *Calumet Region*, 192–94; "Big Standard Oil Works," *Chicago Tribune*, May 11, 1889, 4; Giddens, *Standard Oil*, 13, 16, 20–21, 37; "Greatest in the World," *Chicago Tribune*, September 22, 1889, 9; Swartz, "Whiting Refinery," *Stanolind Record* 4(9): 13–14, 4(10): 14–15; Workers of the Writers' Program of the Work Projects Administration, Indiana (hereafter cited as WPAIN), *The Calumet Region*

Historical Guide, reprint edition (New York: AMS Press, 1939), 230–31; Hidy and Hidy, *Pioneering*, 478; "Frank Barstow Dies on a Train," *New York Times*, August 21, 1909; F. Lawrence Babcock, *The First Fifty: 1889–1939* (Chicago: Standard Oil Company, Indiana, 1939), 10; Edward A. Zivich, *From Zadruga to Oil Refinery: Croation Immigrants and Croatian-Americans in Whiting, Indiana, 1890–1950* (New York: Garland Publishing, 1990), 30.

7. Ball, *Northwestern Indiana*, 405, 408; Swartz, "Whiting Refinery," *Stanolind Record* 4(10): 14–15; Giddens, *Standard Oil*, 20–21; Moore, *Calumet Region*, 186–87.

8. *Whiting Democrat* quoted in Moore, *Calumet Region*, 207. See also "Greatest in the World," *Chicago Tribune*, September 22, 1889, 9; Moore, *Calumet Region*, 194; Giddens, *Standard Oil*, 20–21, 31, 33, 37; Swartz, "Whiting Refinery," *Stanolind Record* 4(10): 14–15; WPAIN, *Historical Guide*, 230.

9. "The Heart of the Octopus," *Chicago Tribune*, June 3, 1906, 34. See also "Standard Oil Gets It," *Chicago Tribune*, June 7, 1890, 9; Thomas H. Cannon, ed., *History of the Lake and Calumet Region of Indiana* (Indianapolis: Historians' Association, 1927), 1:28, 33; William Frederick Howat, ed., *A Standard History of Lake County, Indiana and the Calumet Region* (Chicago: Lewis Publishing, 1915), 651–52, 679; T. H. Ball, *Encyclopedia of Genealogy and Biography of Lake County, Indiana, with a Compendium of History, 1834–1904* (Chicago: Lewis Publishing, 1904).

10. *History Story of Inland Steel Co., 1964*, folder 4, box 1, ISCCC no. 12 (CRA), 2–4.

11. "A Town of 350 Houses Built near Chicago in a Year," *Chicago Tribune*, December 7, 1902, 53. See also Kenneth Schoon, *Calumet Beginnings: Ancient Shorelines and Settlements at the South End of Lake Michigan* (Bloomington: Indiana University Press, 2003), 151; Moore, *Calumet Region*, 223–24.

12. "A Town of 350 Houses Built near Chicago in a Year," *Chicago Tribune*, December 7, 1902, 38; WPAIN, *Historical Guide*, 56, 69–70; Moore, *Calumet Region*, 12.

13. "Fuel Oil Contract," December 4, 1902, folder 13 entitled "Compilation, Historic Tidbits, 1893–1939," box 2, ISCCC no. 12 (CRA); *History Story of Inland Steel Co., 1964*, folder 4, box 1, ISCCC no. 12 (CRA), 17–18; Alfred D. Chandler Jr., *The Visible Hand: The Managerial Revolution in American Business* (Cambridge, MA: Belknap Press, 1977), 422; Ellis, *The Cuyahoga*, 193–97; *History Story of Inland Steel Co., 1964*, folder 4, box 1, ISCCC no. 12 (CRA), 11–12, 24.

14. "Calumet Region Sewage," *Chicago Tribune*, February 16, 1896, 38. See also *Fifty Years of Inland Steel: 1893–1943* (Chicago: Inland Steel Company, 1943); Moore, *Calumet Region*, 131–32; Giddens, *Standard Oil*, 28.

15. William Cronon, *Nature's Metropolis: Chicago and the Great West* (New York: W. W. Norton, 1991), 372. See also Louis P. Cain, "Raising and Watering a

City: Ellis Sylvester Chesbrough and Chicago's First Sanitation System," *Technology and Culture* 13 (1972), 353–72; Louis P. Cain, *Sanitation Strategy for a Lakefront Metropolis: The Case of Chicago* (DeKalb: Northern Illinois University Press, 1978), 69–80.

16. "Calumet Region Sewage," *Chicago Tribune*, February 16, 1896, 38.

17. "A Town of 350 Houses Built near Chicago in a Year," *Chicago Tribune*, December 7, 1902, 53. See also WPAIN, *Historical Guide*, 68–69; Swartz, "Whiting Refinery," *Stanolind Record* 4(9): 11, 12.

18. Harold M. Mayer, "Politics and Land Use: The Indiana Shoreline of Lake Michigan," *Annals of the Association of American Geographers* 54, no. 4 (December 1964), 508–23. See also Moore, *Calumet Region*, 12–14.

19. *History Story of Inland Steel Co.*, folder 4, box 1, ISCCC no. 12 (CRA), ch. 7.

20. *Laws of the State of Indiana, 1907* (Indianapolis: Wm. B. Burford, 1907), 126. The Supreme Court found, according to legal scholar Joseph Sax, the city had no right "to divest itself of authority to govern the whole of an area in which it has responsibility to exercise its police power" by granting "almost the entire waterfront of a major city to a private company." See Joseph L. Sax, "The Public Trust Doctrine in Natural Resource Law: Effective Judicial Intervention," *Michigan Law Review* 68, no. 3 (January 1970): 489–91; *Illinois Central v. State of Illinois*, Supreme Court decision 146 U.S. 387 (1892); Jack H. Archer, Donald L. Connors, Kenneth Laurence, and Robert Bowen, *The Public Trust Doctrine and the Management of America's Coasts* (Amherst: University of Massachusetts Press, 1994), 56–58; Joshua A. T. Salzmann, "Safe Harbor: Chicago's Waterfront and the Political Economy of the Built Environment, 1847–1918" (PhD dissertation, University of Illinois at Chicago, 2008), 66–72; Moore, *Calumet Region*, 100–101; *Fifty Years of Inland Steel*, 11; "Unite to Save Lake Front," *Chicago Tribune*, September 11, 1899, 2. For the more stringent Illinois law see Theodore K. Long, *Report to the Mayor and the City Council of the City of Chicago by the Lake Shore Reclamation Commission* (Chicago: Barnard and Miller, 1912), 304–5; "Case Involves Riparian Rights," *Chicago Tribune*, November 19, 1895, 3.

21. Agnes and John Dvorscak interview, by John Bodnar, May 14, 1991, call number 91-021, "Whiting, Indiana: Generational Memory, 1991–1993," Center for the Study of History and Memory, Indiana University, Bloomington (hereafter cited as CSHM), 17. See also Charles Howard Foulkes, *"Gas!" The Story of the Special Brigade* (London: W. Blackwood and Sons, 1934), 105; Swartz, "Whiting Refinery," *Stanolind Record* 4(11): 12; "How Kerosene Is Made," *Chicago Tribune*, June 21, 1896, 18; Giddens, *Standard Oil*, 26–28; Zivich, *From Zadruga to Oil Refinery*, 38; John J. Marcisz interview, by John Bodnar, March 3, 1992, call number 91-152 (CSHM), 15; John M. Jancosek interview, by David Dabertin, February 9, 1991, call number 91-014 (CSHM), 33.

22. Henry Chandler Cowles, "The Ecological Relations of the Vegetation on the Sand Dunes of Lake Michigan. Part I.—Geographical Relations of the Dune Floras," *Botanical Gazette* 27, no. 2 (February 1899): 110; "The Heart of the Octopus: Whiting, Indiana," *Chicago Tribune*, June 3, 1906, D8; Swartz, "Whiting Refinery," *Stanolind Record* 4(9): 11, 12. See also Joseph Gresko interview, by John Bodnar, May 14, 1991, call number 91-023 (CSHM), 11–13; George Jancosek interview, by David Dabertin, January 28, 1991, call number 91-017 (CSHM), 2.

23. Moore, *Calumet Region*, 108–9, 127.

24. Joseph Novosel Sr. interview, by David Dabertin, October 15, 1990, call number 91-002 (CSHM), 3.

25. Joseph Novosel Sr. interview, by David Dabertin, October 15, 1990, call number 91-002 (CSHM), 31; Joseph Gresko interview, by John Bodnar, May 14, 1991, call number 91-023 (CSHM), 12. See also Delores and William J. Curosh, Michael and Bertha Deluca interview, by John Bodnar, March 2, 1992, call number 91-151 (CSHM), 29–30; Julia and Michael Pukac interview, by John Bodnar, March 2, 1992, call number 91-153 (CSHM), 33–34; Betty Gehrke interview, by John Bodnar, September 28, 1990, call number 91-004 (CSHM), 21; Clarence and Betty Gehrke interview, by John Bodnar, October 11, 1991, call number 91-142 (CSHM), 48; Giddens, *Standard Oil*, 27; Zivich, *From Zadruga*, 7, 25.

26. Arthur E. Gorman, "Survey of Sources of Pollution," *Civil Engineering* 3, no. 9 (September 1933): 519–22; John R. Baylis, "Effect of Certain Industrial Wastes," *Civil Engineering* 3, no. 9 (September 1933): 522–24; "Report of an Investigation of the Pollution of Lake Michigan in the Vicinity of South Chicago and Indiana Harbors," *Public Health Reports* 42, no. 35 (September 2, 1927): 2201. An outbreak of typhoid had visited Chicago in 1891 and killed 1,997, spurring the creation of several more water intake cribs in Lake Michigan farther from the lakeshore. See Heman Spalding and Herman Bundesen, "Control of Typhoid Fever in Chicago," *American Journal of Public Health* 8, no. 5 (May 1918): 359.

27. Maurice Le Bosquet, "Report on Pollution of the Waters of the Grand Calumet River, Little Calumet River, Calumet River, Lake Michigan, Wolf Lake and Their Tributaries," *Proceedings*, vol. 1: *Conference in the Matter of Pollution of the Interstate Waters of the Grand Calumet River, Little Calumet River, Calumet River, Lake Michigan, Wolf Lake and Their Tributaries, March 2, 1965* (Washington DC: US Government Printing Office, 1966; hereafter cited as *CPIW*), 55. See also Moore, *Calumet Region*, 13; Gorman, "Survey of Sources of Pollution," 519; Le Bosquet, "Report on Pollution," 54, 57; "Statement presented by James W. Jardine, Commissioner Department of Water and Sewers, City of Chicago at the Interstate Pollution Conference Held on March 2, 1965," *CPIW*, 380–82. Carolyn G. Shapiro-Shapin, "'A Really Excellent Scientific Contribution': Scientific Creativity, Scientific Professionalism, and the Chicago Drain-

age Case, 1900–1906," *Bulletin of the History of Medicine* 71, no. 3 (Fall 1997); *State of Missouri v. State of Illinois and the Sanitary District of Chicago*, 180 U.S. 208 (1901).

28. James F. Bartuska, "Ozonation at Whiting," *American Water Works Association* 33, no. 11 (November 1941): 2036.

29. James F. Bartuska, "Ozonation at Whiting," 2042. See also "Public Health Service, Drinking Water Standards," rev. 1962, no. 956 (Washington, DC: US Department of Health, Education, and Welfare, 1962), 6; Le Bosquet, "Report on Pollution," 123–24.

30. Cain, *Sanitation Strategy*, 127; Mr. Hyman Gerstein, chief water engineer, Chicago Water Department, *CPIW*, 392–93; Le Bosquet, "Report on Pollution," 83, 130, 198–99; Niagara Falls Geology Facts and Figures, at http://www.niagaraparks .com/media/geology-facts-figures.html/.

31. Although separators were eventually installed so the ballast could be cleaned before dumping, the practice of dumping waste overboard was prevalent in the early twentieth century. Standard Oil Company of New York, Lighterage Department, "Rules and Regulations," May 1, 1910, 5, in box 2.207-H58B, Exxon-Mobil Archives, Briscoe Center for American History, University of Texas at Austin; Joseph Gresko interview, by John Bodnar, May 14, 1991, call number 91-023 (CSHM), 10. See also "Keeping the Water Clean," *Vacuum Oil News* (March 1929): 9; *Great Lakes Basin Framework Study, Appendix 7: Water Quality* (Ann Arbor: Great Lakes Basin Commission, 1975), 88.

32. Frank N. Egerton, "Missed Opportunities: U.S. Fishery Biologists and Productivity of Fish in Green Bay, Saginaw Bay and Western Lake Erie," *Environmental Review* 13, no. 2 (Summer 1989): 50. On Calumet River changes, see Moore, *Calumet Region*, 12. For changes to Lake Michigan's fish population, see *Great Lakes Basin Framework Study, Appendix 8: Fish* (Ann Arbor: Great Lakes Basin Commission, 1975), 5, 19, 17–18.

33. Le Bosquet, "Report on Pollution," 103, also 48, 50, 68, 79, 98, 325. See also Mr. Harold Jordahl Jr., Regional Coordinator, US Department of the Interior, *CPIW*, 334.

34. *Great Lakes Basin Framework Study, Appendix 18: Erosion and Sedimentation* (Ann Arbor: Great Lakes Basin Commission, 1975), 29, 86; William M. Holden, "Hot Water: Menace and Resource," *Science News* 94, no. 7 (August 17, 1968): 164–66; Stanford H. Smith, "Pushed toward Extinction: The Salmon and Trout," in *The Enduring Great Lakes*, ed. John Rousmaniere (New York: W. W. Norton, 1979), 42, 45.

35. "The Alewife Explosion: The 1967 Die-Off in Lake Michigan," *Federal Water Pollution Control Administration, Great Lakes Region* (July 25, 1967): 1.

36. "Alewife Cleanup Program," *New York Times*, February 3, 1968, 26. See also

Jerome Namias, "Nature and Possible Causes of the Northeastern United States Drought during 1962–1965," *Monthly Weather Review* 94, no. 9 (September 1966): 543–54; US Army Corps of Engineers-Detroit District, "Monthly Bulletin of Lake Levels for the Great Lakes" (July 2012): 1, at http://www.lre.usace.army.mil/glhh/; Steinberg, *Down to Earth*, 239; Edward H. Brown Jr., "Population Characteristics and Physical Condition of Alewives, *Alosa Pseudoharengus*, in a Massive Dieoff in Lake Michigan, 1967," *Great Lakes Fishery Commission, Technical Report No. 13* (December 1968); "The Alewife Explosion: The 1967 Die-Off in Lake Michigan," *Federal Water Pollution Control Administration, Great Lakes Region* (July 25, 1967): 1, 13, 15, 31; "Water Pollution Problems of the Great Lakes Area," *Federal Water Pollution Control Administration, Great Lakes Region* (September 1966); "Alewife Control Bill Gains," *New York Times*, March 21, 1968, 94; "Senate Votes Alewife Study," *New York Times*, April 5, 1968, 17.

37. "Pollution Parley Adopts Five Points," *New York Times*, March 13, 1968, 34.

38. *CPIW*, 587–88.

39. Betty Gehrke interview, by John Bodnar, September 28, 1990, call number 91-004 (CSHM), 21; "The Heart of the Octopus: Whiting, Indiana," *Chicago Tribune*, June 3, 1906, D8. For the election of Standard officials, see Moore, *Calumet Region*, 205–8; for tax figures, see WPAIN, *Historical Guide*, 5, and Moore, *Calumet Region*, 207; for Whiting residents on the social quietude achieved through economic dominance, see also Joseph Gresko interview, by John Bodnar, May 14, 1991, call number 91-023 (CSHM), 12.

40. Eugene F. Stoermer, "Bloom and Crash: Algae in the Lakes," in *The Enduring Great Lakes*, ed. John Rousmaniere (New York: W. W. Norton, 1979), 15–16, 18–19. See also Andrew Hurley, *Class, Race, and Industrial Pollution in Gary, Indiana: 1945–1980* (Chapel Hill: University of North Carolina Press, 1995), 34–35; "Prescription for a Dying Lake," *Science News* 93, no. 7 (February 17, 1968): 159; "Dead Fish by the Ton," *Science News* 92, no. 1 (July 1, 1967): 9–10; Holden, "Hot Water," 164–66. Lake Michigan, at ten thousand years old, is geologically young when compared to other lake systems, thus making it more susceptible to disruption from invasive species as the endemic plant and animal species have had less time to acclimate themselves evolutionarily. See Peter H. Gleick, ed., *Water in Crisis, A Guide to the World's Fresh Water Resources* (New York: Oxford University Press, 1993), 120. The rise in stream and lake water temperatures has been documented throughout the Great Lakes, see especially H. A. Regier and W. L. Hartman, "Lake Erie's Fish Community: 150 Years of Cultural Stresses," *Science* 180 (1973): 1248; Robert N. Stavins, "The Problem of the Commons: Still Unsettled after 100 Years," in *Economics of the Environment: Selected Readings*, ed. Robert N. Stavins, 6th ed. (New York: W. W. Norton, 2012), 574.

41. Craig E. Colten, *Industrial Wastes in the Calumet Area, 1869–1970: An His-*

torical Geography, from the Hazardous Waste Research and Information Center, RR-001 (Springfield: Illinois State Museum, 1985), 1; *CPIW*, 112.

42. Michael Gorn, interview with William Ruckelshaus, January 1993, at https://archive.epa.gov/epa/aboutepa/william-d-ruckelshaus-oral-history-interview .html/.

43. Janet Zenke Edwards, *Diana of the Dunes: The True Story of Alice Gray* (Charleston, SC: History Press, 2010), 145. See also Moore, *Calumet Region*, 602–4.

44. Edwards, *Diana of the Dunes*, 48.

45. Moore, *Calumet Region*, 602; Edwards, *Diana of the Dunes*, 142. See also Edwards, *Diana of the Dunes*, 48–49, 87, 92, 96, 98, 101; Isenberg, *Mining California*, 15.

46. Moore, *Calumet Region*, 602; Williamson and Daum, *American Petroleum Industry*, 17.

CONCLUSION

1. Nevins, *Study in Power*, 2:12; Hidy and Hidy, *Pioneering*, 116.

2. "Standard Oil Trust Agreement," folder titled "Standard Oil Company: Founding and Governance Documents: Trust Agreement, 1879," n.d., box 2W30, in ExxonMobil Archives, Briscoe Center for American History, University of Texas at Austin (hereafter ExxonMobil); S. C. T. Dodd to Flagler, "Relative to plan of organizing a Corporation for purpose of holding Stocks of Corporations in various States etc." and S. C. T. Dodd to Flagler, "Opinion relative proposed dissolution of the SOCo as a Corporation existing under the Act of Legislature of the State of Ohio," dated July 23, 1881, in folder titled "Standard Oil Company: Founding and governance documents: S.C.T. Dodd opinions and abstract, 1881," n.d., box 2W30, ExxonMobil. See also Hidy and Hidy, *Pioneering*, 46, 49; August W. Giebelhaus, *Business and Government in the Oil Industry: A Case Study of Sun Oil, 1876–1945* (Greenwich, CT: JAI Press, 1980), 4–5; Leslie D. Manns, "Dominance in the Oil Industry: Standard Oil From 1865 to 1911," in *Market Dominance: How Firms Gain, Hold, or Lose It and the Impact on Economic Performance*, ed. David I. Rosenbaum (Westport, CT: Praeger, 1998), 20. Joel Bakan has shown that following the *Dodge v. Ford* decision in 1916, corporate owners were legally bound to prioritize investor profit before any other consideration. Joel Bakan, *The Corporation: The Pathological Pursuit of Profit and Power* (New York: Free Press, 2004), 36–37.

3. Hidy and Hidy, *Pioneering*, 308. Ralph and Muriel Hidy found that New Jersey granted an average of 2,172 corporate charters annually between the years of 1899 and 1901 alone. Hidy and Hidy, *Pioneering*, 309. See also Giebelhaus, *Business and Government*, 5; Horwitz, *Transformation, 1870–1960*, 84.

4. H. J. Haynes, *Standard Oil Company of California: 100 Years Helping to*

Create the Future (New York: The Newcomen Society in North America, 1980), 9; Hidy and Hidy, *Pioneering*, 42, 126, 151. For a modern look at Standard Oil's overseas empire and its relationship to US political goals, see Alison Fleig Frank, *Oil Empire: Visions of Prosperity in Austrian Galicia* (Cambridge, MA: Harvard University Press, 2005).

5. William Libby to Charles Pratt and Co., New York, April 9, 1879, box 2.207/ J24, ExxonMobil. See also F. S. Cooper Jr., "The Flying Red Horse in Japan" (notes), June 2, 1961, folder titled "Asia, Japan, 1879–1993," box 2.207-E174, ExxonMobil.

6. Horwitz, *Transformation, 1870–1960*, 80–93.

7. Hidy and Hidy, *Pioneering*, 452.

8. Hidy and Hidy, *Pioneering*, 301. See also Yergin, *The Prize*, 60–63, 132, 153.

9. Yergin, *The Prize*, 84–87.

10. "Supreme Court Rules Standard Oil Company Is Illegal Trust," *Los Angeles Times*, May 16, 1911. See also *William Macey vs. The Standard Oil Company*, August 5, 1872, in *Annals of Cleveland Court Record Series*, vol. 8: *1871–1872* (Cleveland, 1939): 263.

11. Richard Hofstadter, *The Age of Reform: From Bryan to F.D.R.* (New York: Alfred A. Knopf, 1989), 248.

12. Yergin, *The Prize*, 113; "Financier's Fortune in Oil Amassed in Industrial Era of 'Rugged Individualism,'" *New York Times*, May 24, 1937; A. D. Kaplan, *Big Enterprise in a Competitive System* (Washington, DC: The Brookings Institution, 1964), 144.

13. "The Exxon-Mobil Merger: Hearings before the Subcommittee on Energy and Power of the Committee on Commerce," House of Representatives, 106th Congress, 1st sess., March 10, 11, 1999, vol. 4 (Washington, DC: US Government Printing Office, 1999). Anthony Sampson, *The Seven Sisters: The Great Oil Companies and the World They Shaped* (New York: Viking Press, 1975); Adolf A. Berle Jr., and Gardiner C. Means, *The Corporation and Private Property* (New York: MacMillan, 1934), 45, 46.

14. Ivy Lee to JDR, February 11, 1915, folder 3, 7, box 4, Record Group 3, RAC; W. O. Inglis to JDR, July 8, 1925, folder 108, box 15, Record Group 2, RAC. See also JDR, *Random Reminiscences*; Hawke, "Interview," 2, 6–7. Inglis salary memo, September 6, 1927, Inglis Notes, folder 3, box 4, Record Group 3, and W. O. Inglis to JDR, July 8, 1925, folder 108, box 15, Record Group 2, RAC.

15. Ernst, *"Dear Father,"* 201. See also Edward N. Akin, *Flagler* (Gainesville: University Press of Florida, 1992), xi. In a review of contemporary scholarship dealing with business history, Oscar Handlin, while allowing that Nevins "On points of detail . . . is thoroughly reliable," argued that "the work is essentially apologetic." In the review, Handlin lumps Nevins's *Study in Power* with F. A. Hayek's anthology *Capitalism and the Historians* in asking "the wrong questions"

of "whether the captains of industry were moral or not." Instead, he found Nevins and Hayek equally guilty of "obscur[ing] the much more vital issues of the means by which 'capitalism' transformed the economy, the forms it took, and its effects upon the whole society." Oscar Handlin, "Capitalism, Power, and the Historians: An Essay Review," *New England Quarterly* 28, no. 1 (March 1955): 99–107. See also Ray Eldon Hiebert, *Courtier to the Crowd: The Story of Ivy Lee and the Development of Public Relations* (Ames: Iowa State University Press, 1966), frontispiece, xi, 97, 104, 114–15, 119, 131–34; Thomas G. Andrews, *Killing for Coal: America's Deadliest Labor War* (Cambridge, MA: Harvard University Press, 2008), 8.

16. *Activities of the Refinery Loss Committee: Baytown Refinery* (Baytown, TX: Humble Oil and Refining Company, 1952), in ExxonMobil Archives.

17. "Into Thin Air," *The Lamp* (June–September 1948), Standard Oil Company (New Jersey), New York, in ExxonMobil Archives.

18. "Press Relations for the Mobil Traveler," folder entitled "Mobil International Division: General," box 2.207-G115; "Instructions for the Rating of Professional Laboratory Employees," in folder entitled "Human Resources: Employee Relations, R&D Supervisor's Manual, c.a. 1950," box 2.207-E39, ExxonMobil.

19. Alfred Runte, *National Parks: The American Experience*, 4th ed. (New York: Taylor Trade Publishing, 2010), 105, 107, 114, 128–30, 134; Dyan Zaslowsky and T. H. Watkins, *These American Lands: Parks, Wilderness, and the Public Lands* (Washington, DC: Island Press, 1994), 30–31. Albright, while committed to the concept of ecology, was an ardent supporter of the government's predator eradication program. See Dan Flores, *Coyote America: A Natural and Supernatural History* (New York: Basic Books, 2016), 138.

20. "Exxon Exec Says Doesn't Know Montana Spill's Cause," The Wall Street Journal Digital Network, July 14, 2011, at http://www.marketwatch.com/story /exxon-exec-says-doesnt-know-montana-spills-cause-2011-07-14?reflink=MW_news _stmp/. "Exxon Must Pay $1m for 2011 Spill into Montana's Yellowstone River," The Guardian News and Media Ltd., June 12, 2015, at http://www.theguardian .com/environment/2015/jun/13/exxon-must-pay-1m-for-2011-spill-into-montanas -yellowstone-river#img-1/; Norman Maclean, *A River Runs through It and Other Stories* (Chicago: University of Chicago Press, 1976).

21. Keeter from T. A. Frail, "What Will America Look like in 2050?" *Smithsonian Magazine*, August 2010, at http://www.smithsonianmag.com/innovation /poll-americans-predict-life-in-2050-548250/; Steven Stoll, *The Great Delusion: A Mad Inventor, Death in the Tropics, and the Utopian Origins of Economic Growth* (New York: Hill and Wang, 2008), 165.

BIBLIOGRAPHY

MANUSCRIPT COLLECTIONS

Briscoe Center for American History, University of Texas at Austin.
    ExxonMobil Archives.
Center for the Study of History and Memory (CSHM), Indiana University, Bloomington.
    Whiting, Indiana: Generational Memory, 1991–1993.
Cleveland Public Library, Cleveland, Ohio.
    Special Collections.
Indiana University Northwest, Gary, Indiana.
    Calumet Regional Archives (CRA).
Library of Congress, Washington, DC.
    Early Motion Pictures Collection, 1897–1920, Edison Films Catalog.
Rockefeller Archive Center (RAC), Sleepy Hollow, New York.
    Rockefeller Family Archives.
    Record Group 3. Inglis Notes.
University of Chicago Library, Chicago, Illinois.
    Map Collection.
Western Reserve Historical Society(WRHS), Cleveland, Ohio.
    Annals of Cleveland Court Record Series. Vol. 8, 1871–1872. Vol. 10, 1875–1877.
    Special Collections.

NEWSPAPERS AND JOURNALS

Bayonne Times
Chicago Sun-Times
Chicago Tribune
Cleveland Leader
Cleveland Morning Leader
Cleveland Plain Dealer
Cleveland Press
Cleveland Recorder
Cleveland World
Journal of the Senate of Minnesota
The Lamp
Los Angeles Times

Michigan Law Review
New York Times
Ohio Public Health Journal
Oil Paint and Drug Reporter
Providence, R.I. Sunday Journal
Vacuum Oil News
Whiting Democrat

PRIMARY AND SECONDARY SOURCES

Adler, Jonathan, ed. *Ecology, Liberty and Property: A Free Market Environmental Reader*. Washington, DC: Competitive Enterprise Institute, 2000.

Adler, Jonathan. "Fables of the Cuyahoga: Reconstructing a History of Environmental Protection." *Fordham Environmental Law Journal* 14, no. 89 (2002): 89–146.

Akin, Edward N. *Flagler*. Gainesville: University Press of Florida, 1992.

Allen, James B. *The Company Town in the American West*. Norman: University of Oklahoma Press, 1966.

Andrews, Thomas G. *Killing for Coal: America's Deadliest Labor War*. Cambridge, MA: Harvard University Press, 2008.

Applebome, Peter. "They Used to Say Whale Oil Was Indispensable, Too." *New York Times*, August 3, 2008.

Appleby, Joyce. *Capitalism and a New Social Order: The Republican Vision of the 1790s*. New York: New York University Press, 1984.

Armistead, George, Jr. *Safety in Petroleum Refining and Related Industries*. New York: John G. Simmonds, 1950.

Ashworth, William. *The Late, Great Lakes: An Environmental History*. New York: Alfred A. Knopf, 1986.

Babcock, F. Lawrence. *The First Fifty, 1889–1939*. Chicago: Standard Oil Company (Indiana), 1939.

Bakan, Joel. *The Corporation: The Pathological Pursuit of Profit and Power*. New York: Free Press, 2004.

Ball, T. H. *Northwestern Indiana: From 1800 to 1900*. Chicago: Donohue and Henneberry, 1900.

Barrow, William C. "The Euclid Heights Allotment." MA thesis, Cornell University, Ithaca, New York, 1996.

Baylis, John R. "Effect of Certain Industrial Wastes." *Civil Engineering* 3, no. 9 (September 1933): 522–24.

Beard, Charles A., and Mary R. Beard. *A Basic History of the United States*. New York: Doubleday, Doran, 1944.

Bedford, Henry F., and Trevor Colbourn. *The Americans: A Brief History*. 2nd ed. New York: Harcourt, Brace, Jovanovich, 1976.

Bellamy, John Stark. *The Killer in the Attic: And More True Tales of Crime and Disaster from Cleveland's Past*. Cleveland, OH: Gray, 2002.

Benidickson, Jamie. *The Culture of Flushing*. Vancouver: UBC Press, 2007.

Berle, Adolf A., Jr., and Gardiner C. Means. *The Corporation and Private Property*. New York: MacMillan, 1934.

Black, Brian. *Petrolia: The Landscape of America's First Oil Boom*. Baltimore: Johns Hopkins University Press, 2000.

Bogart, Ernest Ludlow. *Internal Improvements and State Debt in Ohio: An Essay in Economic History*. New York: Longmans, Green, 1924.

The Book of Clevelanders: A Biographical Dictionary of Living Men of the City of Cleveland. Cleveland: Burrows Bros., 1914.

Bottomore, Tom, ed. *A Dictionary of Marxist Thought*. Cambridge, MA: Harvard University Press, 1983.

Brands, H. W. *American Colossus: The Triumph of Capitalism, 1865–1900*. New York: Doubleday, 2010.

Brown, Richard D. "Modernization and the Modern Personality in Early America, 1600–1865: A Sketch of a Synthesis." *Journal of Interdisciplinary History* 2, no. 3 (Winter 1972): 201–28.

Cain, Louis P. *Sanitation Strategy for a Lakefront Metropolis: The Case of Chicago*. DeKalb: Northern Illinois University Press, 1978.

Calloway, Colin G. *The American Revolution in Indian Country: Crisis and Diversity in Native American Communities*. Cambridge: Cambridge University Press, 1995.

Campbell-Mohn, Celia, ed. *Environmental Law: From Resources to Recovery*. St. Paul: West Publishing, 1993.

Carr, Albert Z. *John D. Rockefeller's Secret Weapon*. New York: McGraw-Hill, 1962.

Carter, Luther J. "Lake Michigan: Salmon Help to Redress the Balance." *Science*, n.s. 161, no. 3841 (August 9, 1968): 551–55.

Chandler, Alfred D., Jr., ed. *The Coming of Managerial Capitalism*. Homewood, IL: Richard D. Irwin, 1985.

Chandler, Alfred D., Jr. "The Standard Oil Company—Combination, Consolidation, and Integration." In *The Coming of Managerial Capitalism*, edited by Alfred D. Chandler Jr., 342–71. Homewood, IL: Richard D. Irwin, 1985.

Chandler, Alfred D., Jr. *The Visible Hand: The Managerial Revolution in American Business*. Cambridge, MA: Belknap Press, 1977.

Chapman, Edmund H. *Cleveland: Village to Metropolis*. Cleveland: The Press of Western Reserve University, 1964.

Chapman, Ralph E., and George G. Cummings. "A Study of the 'Sootfall' in the City of Cleveland, Ohio." BS thesis, Case School of Applied Science, Cleveland, Ohio, 1924.

Chernow, Ron. *Titan: The Life of John D. Rockefeller, Sr.* New York: Random House, 1998.

Chester, Robert. "Manufacturing Danger: The Perils of Place." PhD thesis, University of California, Davis, 2009.

Chrislock, Winston. "Cleveland's Czechs." In *Identity, Conflict, and Cooperation: Central Europeans in Cleveland, 1850–1930,* edited by David C. Hammack, Diane L. Grabowski, and John J. Grabowski, 12–43. Cleveland: Western Reserve Historical Society, 2002.

Clarke, K. C., and Jeffrey J. Hemphill. "The Santa Barbara Oil Spill: A Retrospective." *Yearbook of the Association of Pacific Coast Geographers,* vol. 64, edited by Darrick Danta, 157–62. Honolulu: University of Hawai'i Press, 2002.

Cleveland, Ohio. *Reports of the Departments of the Government of the City of Cleveland for the Year Ending December 31, 1864.* Cleveland: Fairbanks, Benedict, 1865.

Cleveland, Ohio. *Reports of the Departments of the Government of the City of Cleveland for the Year Ending December 31, 1865.* Cleveland: Leader Company, 1866.

Cleveland, Ohio. *Reports of the Departments of the Government of the City of Cleveland for the Year Ending December 31, 1866.* Cleveland: Fairbanks, Benedict, 1867.

Cleveland, Ohio. *Reports of the Departments of the Government of the City of Cleveland for the Year Ending December 31, 1868.* Cleveland: Leader Book and Job Office, 1869.

Cleveland, Ohio. *Reports of the Departments of the Government of the City of Cleveland for the Year Ending December 31, 1871.* Cleveland: Leader Book and Job Office, 1872.

Cleveland, Ohio. *Reports of the Departments of the Government of the City of Cleveland for the Year Ending December 31, 1872.* Cleveland: Waechter am Erie Printing, 1873.

Cleveland, Ohio. *Reports of the Departments of the Government of the City of Cleveland for the Year Ending December 31, 1873.* Cleveland: Fairbanks, Benedict, 1874.

Cleveland, Ohio. *Reports of the Departments of the Government of the City of Cleveland for the Year Ending December 31, 1874.* Cleveland: Co-operative Printing, 1875.

Cleveland, Ohio. *Reports of the Departments of the Government of the City of Cleveland for the Year Ending December 31, 1875.* Cleveland: Co-operative Printing, 1876.

Cleveland, Ohio. *Reports of the Departments of the Government of the City of Cleveland for the Year Ending December 31, 1876.* Cleveland: A. W. Fairbanks, 1877.

Cleveland, Ohio. *Reports of the Departments of the Government of the City of Cleveland for the Year Ending December 31, 1878.* Cleveland: Wiseman and Harvey, 1879.

Cleveland, Ohio. *Reports of the Departments of the Government of the City of Cleveland for the Year Ending December 31, 1879.* Cleveland: Wiseman and Harvey, 1880.

Cleveland, Ohio. *Reports of the Departments of the Government of the City of Cleveland for the Year Ending December 31, 1880.* Cleveland: Home Companion Publishing, 1881.

Cleveland, Ohio. *Reports of the Departments of the Government of the City of Cleveland for the Year Ending December 31, 1881.* Cleveland: Home Companion Publishing, 1882.

Cleveland, Ohio. *Reports of the Departments of the Government of the City of Cleveland for the Year Ending December 31, 1884.* Cleveland: Peerless Printing, 1885.

Cleveland, Ohio. *Reports of the Departments of the Government of the City of Cleveland for the Year Ending December 31, 1886.* Cleveland: Peerless Printing, 1887.

Cleveland, Ohio. *Reports of the Departments of the Government of the City of Cleveland for the Year Ending December 31, 1887.* Cleveland: Cleveland Printing and Publishing, 1888.

Cleveland, Ohio. *Reports of the Departments of the Government of the City of Cleveland for the Year Ending December 31, 1890.* Cleveland: Cleveland Printing and Publishing, 1891.

Cleveland, Ohio. *Reports of the Departments of the Government of the City of Cleveland for the Year Ending December 31, 1891.* Cleveland: Cleveland Printing and Publishing, 1892.

Cleveland, Ohio. *Reports of the Departments of the Government of the City of Cleveland for the Year Ending December 31, 1892.* Cleveland: Cleveland Printing and Publishing, 1893.

Cleveland, Ohio. *Reports of the Departments of the Government of the City of Cleveland for the Year Ending December 31, 1893.* Cleveland: J. B. Savage, Print, 1894.

Cleveland, Ohio. *Reports of the Departments of the Government of the City of Cleveland for the Year Ending December 31, 1895.* Cleveland: Brooks, 1896.

Cleveland, Ohio. *Reports of the Departments of the Government of the City of Cleveland for the Year Ending December 31, 1896.* Cleveland: Cleveland Printing and Publishing, 1897.

Cleveland, Ohio. *Reports of the Departments of the Government of the City of Cleveland for the Year Ending December 31, 1897.* Cleveland: Brooks, 1898.

Cleveland, Ohio. *Reports of the Departments of the Government of the City of Cleveland for the Year Ending December 31, 1898.* Cleveland: Cleveland Printing and Publishing, 1899.

Cleveland, Ohio. *Reports of the Departments of the Government of the City of Cleveland for the Year Ending December 31, 1902.* Cleveland: A. S. Gilman Printing, 1903.

Cobleigh, N. S. *The Manufactures, Trade, and Commerce of Cleveland, 1880–81.* Cleveland, OH: Short and Forman, 1881.

Coleman, Jon T. *Vicious: Wolves and Men in America.* New Haven, CT: Yale University Press, 2004.

Coleridge, Samuel Taylor. *The Rime of the Ancient Mariner.* New York: Harper and Bros., 1877.

Cowles, Henry Chandler. "The Ecological Relations of the Vegetation on the Sand Dunes of Lake Michigan. Part I.—Geographical Relations of the Dune Floras." *Botanical Gazette* 27, no. 2 (February 1899): 95–117.

Crohurst, H. R., and M. V. Veldee. "Report of an Investigation of the Pollution of Lake Michigan in the Vicinity of South Chicago and Indiana Harbors." *Public Health Reports* 42, no. 35 (September 2, 1927): 2200–202.

Cronon, William. *Changes in the Land: Indians, Colonists, and the Ecology of New England.* New York: Hill and Wang, 1983.

Cronon, William. *Nature's Metropolis: Chicago and the Great West.* New York: W. W. Norton, 1991.

Davis, Lance E., Robert E. Gallman, and Karin Gleiter. *In Pursuit of Leviathan: Technology, Institutions, Productivity, and Profits in American Whaling, 1816–1906.* Chicago: University of Chicago Press, 1997.

"Dead Fish by the Ton." *Science News* 92, no. 1 (July 1, 1967): 9–10.

Diamond, Jared. *Collapse: How Societies Choose to Fail or Succeed.* New York: Viking, 2005.

Diller, Oliver D. *Ohio's Forest Resources: Progress Report Based on a Survey Conducted during 1939–1943 and a Recommended Long-Range Forestry Program for Ohio.* Agricultural Experiment Station, Wooster. Ohio Department of Forestry Publication no. 76 (1944).

Dodd, S. C. T. *Trusts.* New York: self-published, 1900.

Dutka, Alan. *Cleveland Calamities: A History of Storm, Fire, and Pestilence.* Charleston, SC: History Press, 2014.

Edwards, Janet Zenke. *Diana of the Dunes: The True Story of Alice Gray.* Charleston, SC: History Press, 2010.

Ellis, William Donohue. *The Cuyahoga.* Dayton, OH: Landfall Press, 1975.

Elmore, Bartow J. *Citizen Coke: The Making of Coca-Cola Capitalism*. New York: W. W. Norton, 2015.

Elwood-Akers, Virginia. *Caroline Severance*. New York: iUniverse, 2010.

Ernst, Joseph W., ed. *"Dear Father"/"Dear Son": Correspondence of John D. Rockefeller and John D. Rockefeller, Jr*. New York: Fordham University Press, 1994.

Evans, Robert. "Blast from the Past." *Smithsonian* (July 2002). At http://www.smithsonianmag.com/history/blast-from-the-past-65102374/.

"The Exxon-Mobil Merger: Hearings before the Subcommittee on Energy and Power of the Committee on Commerce." House of Representatives, 106th Congress, first session, March 10, 11, 1999. Vol. 4. Washington, DC: US Government Printing Office, 1999.

Fagan, Brian. *The Little Ice Age: How Climate Made History, 1300–1850*. New York: Basic Books, 2000.

Fifty Years of Inland Steel, 1893–1943. Chicago: Inland Steel Company, 1943.

Flores, Dan. "Bison Ecology and Bison Diplomacy: The Southern Plains from 1800 to 1850." *Journal of American History* 78 (1991): 465–85.

Flores, Dan. *Coyote America: A Natural and Supernatural History*. New York: Basic Books, 2016.

Flynn, John T. *God's Gold: The Story of Rockefeller and His Times*. New York: Harcourt, Brace, 1932.

Foner, Eric. *Free Soil, Free Labor, Free Men: The Ideology of the Republican Party before the Civil War*. New York: Oxford University Press, 1970.

Foner, Eric. *The Story of American Freedom*. New York: W. W. Norton, 1998.

Forbes, R. J. *More Studies in Early Petroleum History, 1860–1880*. Leiden: E. J. Brill, 1959.

Foster, John Bellamy. *Marx's Ecology: Materialism and Nature*. New York: Monthly Review Press, 2000.

Foulkes, Charles Howard. *"Gas!" The Story of the Special Brigade*. London: W. Blackwood and Sons, 1934.

Frank, Alison Fleig. *Oil Empire: Visions of Prosperity in Austrian Galicia*. Cambridge, MA: Harvard University Press, 2005.

Frank, Alison Fleig. "The Petroleum War of 1910: Standard Oil, Austria, and the Limits of the Multinational Corporation." *American Historical Review* 114, no. 1 (2009): 16–41.

Friedman, Walter A. *Birth of a Salesman: The Transformation of Selling in America*. Cambridge, MA: Harvard University Press, 2004.

Gesner, Abraham. *A Practical Treatise on Coal, Petroleum and Other Distilled Oils*. 2nd ed. New York: Augustus M. Kelley, 1968.

Giddens, Paul H. *Standard Oil Company (Indiana): Oil Pioneer of the Middle West*. New York: Appleton-Century-Crofts, 1955.

Giebelhaus, August W. *Business and Government in the Oil Industry: A Case Study of Sun Oil, 1876–1945*. Greenwich, CT: JAI Press, 1980.

Gleick, Peter H., ed. *Water in Crisis: A Guide to the World's Fresh Water Resources*. New York: Oxford University Press, 1993.

Gorman, Arthur E. "Survey of Sources of Pollution." *Civil Engineering* 3, no. 9 (September 1933): 519–22.

Gorman, Hugh S. *Redefining Efficiency: Pollution Concerns, Regulatory Mechanisms, and Technological Change in the U.S. Petroleum Industry*. Akron: University of Akron Press, 2001.

Gottlieb, Robert. *Forcing the Spring: The Transformation of the American Environmental Movement*. Washington, DC: Island Press, 1993.

Grandin, Greg. *Fordlandia: The Rise and Fall of Henry Ford's Forgotten Jungle City*. New York: Metropolitan Books, 2009.

Great Lakes Basin Framework Study, Appendix 7: Water Quality. Ann Arbor: Great Lakes Basin Commission, 1975.

Great Lakes Basin Framework Study, Appendix 8: Fish. Ann Arbor: Great Lakes Basin Commission, 1975.

Great Lakes Basin Framework Study, Appendix 18: Erosion and Sedimentation. Ann Arbor: Great Lakes Basin Commission, 1975.

Grinder, R. Dale. "The Battle for Clean Air: The Smoke Problem in Post–Civil War America." In *Pollution and Reform in American Cities, 1870–1930*, edited by Martin V. Melosi, 83–103. Austin: University of Texas Press, 1980.

Gugliotta, Angela. "'Hell with the Lid Taken Off': A Cultural History of Air Pollution—Pittsburgh." PhD dissertation, University of Notre Dame, Notre Dame, Indiana, 2004.

Habermas, Jürgen. *Toward a Rational Society: Student Protest, Science, and Politics*. Boston: Beacon Press, 1970.

Handlin, Oscar. "Capitalism, Power, and the Historians: An Essay Review." *New England Quarterly* 28, no. 1 (March 1955): 99–107.

Harvey, David. *The Condition of Postmodernity: An Enquiry into the Origins of Cultural Change*. Cambridge, UK: Blackwell, 1989.

Hatcher, Harlan. *The Western Reserve: The Story of New Connecticut in Ohio*. Cleveland, OH: World Publishing, 1966.

Hawke, David Freeman. "John D. Rockefeller Interview: 1917–1920." Microfilm. Sleep Hollow, NY: Rockefeller Archive Center, 1984.

Haynes, H. J. *Standard Oil Company of California: 100 Years Helping to Create the Future*. New York: The Newcomen Society in North America, 1980.

Hazen, Margaret Hindle, and Robert M. Hazen. *Keepers of the Flame: The Role of Fire in American Culture, 1775–1925*. Princeton, NJ: Princeton University Press, 1992.

Heimann, Harry. "Effects of Air Pollution on Human Health." In World Health Organization, *Air Pollution*, 159–220. New York: Columbia University Press, 1961.

Henry, J. T. *The Early and Later History of Petroleum*. Vol. 1. New York: Burt Franklin, 1873.

Hidy, Ralph W., and Muriel E. Hidy. *Pioneering in Big Business, 1882–1911*. Vol. 1. New York: Harper and Brothers, 1955.

Hiebert, Ray Eldon. *Courtier to the Crowd: The Story of Ivy Lee and the Development of Public Relations*. Ames: Iowa State University Press, 1966.

Hofstadter, Richard. *The Age of Reform: From Bryan to F.D.R.* New York: Alfred A. Knopf, 1989.

Holden, William M. "Hot Water: Menace and Resource." *Science News* 94, no. 7 (August 17, 1968): 164–66.

Horwitz, Morton J. *The Transformation of American Law, 1780–1860*. Cambridge, MA: Harvard University Press, 1976.

Horwitz, Morton J. *The Transformation of American Law, 1870–1960: The Crisis of Legal Orthodoxy*. New York: Oxford University Press, 1992.

Hoyer, Ernest F. "Yesterday and Today: A Short History of Constable Hook." *The Messenger*, December 24, 1920. Standard Oil Company, New Jersey.

Hughes, Thomas P. "Technological Momentum." In *Does Technology Drive History? The Dilemma of Technological Determinism*, edited by Leo Marx and Merritt Roe Smith, 101–13. Cambridge: MIT Press, 1994.

Hurley, Andrew. *Class, Race, and Industrial Pollution in Gary, Indiana, 1945–1980*. Chapel Hill: University of North Carolina Press, 1995.

Hurley, Andrew. "Creating Ecological Wastelands: Oil Pollution in New York City, 1870–1900." *Journal of Urban History* 20 (May 1994): 340–64.

Hurst, James Willard. *Law and the Conditions of Freedom in the Nineteenth-Century United States*. Madison: University of Wisconsin Press, 1956.

Isenberg, Andrew C. *The Destruction of the Bison: An Environmental History, 1750–1920*. Cambridge: Cambridge University Press, 2000.

Isenberg, Andrew C. *Mining California: An Ecological History*. New York: Hill and Wang, 2005.

Jackson, Daniel D. *Report on the Sanitary Condition of the Cleveland Water Supply on the Probable Effect of the Proposed Changes in Sewage Disposal and on the Various Sources of Typhoid Fever in Cleveland*. Cleveland, OH: City of Cleveland, 1912.

Jackson, Kenneth T. *Crabgrass Frontier: The Suburbanization of the United States*. New York: Oxford University Press, 1985.

Jacoby, Karl. *Crimes against Nature: Squatters, Poachers, Thieves, and the Hidden History of the American Conservation Movement*. Berkeley: University of California Press, 2003.

Japour, Maxcine J. *Petroleum Refining and Manufacturing Processes*. Los Angeles: Wetzel Publishing, 1939.

Jefferson, Thomas. *Notes on the State of Virginia*. Richmond: J. W. Randolph, 1853.

Jenkins, J. T. *A History of the Whale Fisheries*. London: H. F. and G. Witherby, 1921.

Jones, Christopher F. *Routes of Power: Energy and Modern America*. Cambridge, MA: Harvard University Press, 2014.

Kaplan, A. D. *Big Enterprise in a Competitive System*. Washington, DC: The Brookings Institution, 1964.

Kasson, John F. "Republican Values as a Dynamic Factor." In *Problems in American Civilization: The Industrial Revolution*, edited by Gary J. Kornblith, 3–12. New York: Houghton Mifflin, 1998.

Kennedy, James Harrison. *A History of the City of Cleveland: Its Settlement, Rise, and Progress, 1796–1896*. Cleveland: Imperial Press, 1896.

Klingle, Matthew. *Emerald City: An Environmental History of Seattle*. New Haven, CT: Yale University Press, 2007.

Lafferty, Michael B., ed. *Ohio's Natural Heritage*. Columbus: Ohio Academy of Science, 1979.

"Lamprey on the Run." *Wisconsin Conservation Bulletin* 29–30 (1964).

Laws of the State of Indiana, 1907. Indianapolis: Wm. B. Burford, 1907.

Le Bosquet, Maurice. "Report on Pollution of the Waters of the Grand Calumet River, Little Calumet River, Calumet River, Lake Michigan, Wolf Lake and Their Tributaries." In *Proceedings*, vol. 1: *Conference in the Matter of Pollution of the Interstate Waters of the Grand Calumet River, Little Calumet River, Calumet River, Lake Michigan, Wolf Lake and Their Tributaries* (March 2, 1965), 43–163. Chicago. Washington, DC: US Government Printing Office, 1966.

LeCain, Timothy J. *Mass Destruction: The Men and Giant Mines that Wired America and Scarred the Planet*. New Brunswick, NJ: Rutgers University Press, 2009.

Lennon, Robert E. "Control of Freshwater Fish with Chemicals." *Proceedings of the Fourth Vertebrate Pest Conference*. West Sacramento, California, 1970, sponsored by the California Vertebrate Pest Committee.

Limerick, Patricia Nelson. *The Legacy of Conquest: The Unbroken Past of the American West*. New York: W. W. Norton, 1987.

Linder, Marc, and Lawrence S. Zacharias. *Of Cabbages and Kings County: Agriculture and the Formation of Modern Brooklyn*. Iowa City: University of Iowa Press, 1999.

Livesay, Harold C. "From Steeples to Smokestacks: The Birth of the Modern Corporation in Cleveland." In *The Birth of Modern Cleveland, 1865–1930*, edited by Thomas F. Campbell and Edward M. Miggins, 24–71. Cleveland, OH: Western Reserve Historical Society, 1988.

Lloyd, Henry Demarest. *Wealth Against Commonwealth*. New York: Harper and Brothers, 1894.

Lomax, John A., and Alan Lomax. *Cowboy Songs and Other Frontier Ballads*. New York: Macmillan, 1948.

Long, Theodore K. *Report to the Mayor and the City Council of the City of Chicago by the Lake Shore Reclamation Commission*. Chicago: Barnard and Miller, 1912.

Lubetkin, M. John. *Jay Cooke's Gamble: The Northern Pacific Railroad, the Sioux, and the Panic of 1873*. Norman: University of Oklahoma Press, 2006.

Lucier, Paul. "The Professional and the Scientist in Nineteenth-Century America." *Isis* 100 (2009): 699–732.

Macey, David, ed. *The Penguin Dictionary of Critical Theory*. London: Penguin Books, 2000.

Maclean, Norman. *A River Runs Through It and Other Stories*. Chicago: University of Chicago Press, 1976.

Manning, Thomas G. *The Standard Oil Company: The Rise of a National Monopoly*. New York: Holt, Rinehart and Winston, 1960.

Marx, Karl. *Capital: A Critique of Political Economy*. Vol. 1. Translated by Ben Fowkes. New York: Penguin Books, 1990.

Mayer, Harold M. "Politics and Land Use: The Indiana Shoreline of Lake Michigan." *Annals of the Association of American Geographers* 54, no. 4 (December 1964): 508–23.

McCoy, Drew R. *The Elusive Republic: Political Economy in Jeffersonian America*. Chapel Hill: University of North Carolina Press, 1980.

McEvoy, Arthur F. *The Fisherman's Problem: Ecology and Law in the California Fisheries, 1850–1980*. New York: Cambridge University Press, 1986.

McNeill, J. R. *Something New under the Sun: An Environmental History of the Twentieth-Century World*. New York: Norton, 2000.

Meadows, Donella, Jorgen Randers, and Dennis Meadows. *Limits to Growth: The Thirty-Year Update*. White River Junction, VT: Chelsea Green Publishing, 2004.

Melosi, Martin V. *The Sanitary City*. Baltimore: Johns Hopkins University Press, 2000.

Melville, Herman. *Moby Dick; or, The White Whale*. Boston: C. H. Simonds, 1922.

Miller, Carol Poh. *Cleveland Metroparks, Past and Present*. Cleveland: Cleveland Metroparks, 1992.

Milner II, Clyde A., Patricia Nelson Limerick, and Charles E. Rankin, eds. *Trails: Toward a New Western History*. Lawrence: University of Kansas Press, 1991.

Montague, Gilbert Holland. "The Later History of the Standard Oil Company." *Quarterly Journal of Economics* 17, no. 2 (February 1903): 293–325.

Moodys Manual of Railroads and Corporation Securities. Vol. 2, *Industrial Section.* New York: Poor's Publishing, 1921.

Moore, Powell A. *The Calumet Region: Indiana's Last Frontier.* Indianapolis: Indiana Historical Bureau, 1959.

Nevins, Allan. *John D. Rockefeller: The Heroic Age of American Enterprise.* 2 vols. New York: Charles Scribner's Sons, 1940.

Nevins, Allan. *Study in Power: John D. Rockefeller, Industrialist and Philanthropist.* 2 vols. New York: Charles Scribner's Sons, 1953.

Nicolson, Marjorie Hope. *Mountain Gloom and Mountain Glory: The Development of the Aesthetics of the Infinite.* New York: W. W. Norton, 1963.

Orth, Samuel. *A History of Cleveland.* Vol. 1. Chicago: S. J. Clarke Publishing, 1910.

Peckham, Stephen Farnum. *Report on the Production, Technology, and Uses of Petroleum and Its Products.* Washington, DC: Government Printing Office, Department of the Interior, 1885.

Persky, Joseph. "Retrospectives: The Ethology of Homo Economicus." *Journal of Economic Perspectives* 9, no. 2 (Spring 1995): 221–31.

Phillips, Leigh. *Austerity Ecology and the Collapse-Porn Addicts: A Defense of Growth, Progress, Industry and Stuff.* Croydon, UK: Zero Books, 2015.

Polyani, Karl. *The Great Transformation: The Political and Economic Origins of Our Time.* 2nd ed. Boston: Beacon Press, 2001.

Porter, Philip W. *Cleveland: Confused City on a Seesaw.* Columbus: Ohio State University Press, 1976.

Pratt, R. Winthrop. "A New Move for Water Filtration at Cleveland, Ohio." *Engineering News* 69, no. 21 (May 22, 1913).

Pratt, R. Winthrop. "Sewage Disposal Investigations at Cleveland." *Engineering News* 69, no. 7 (February 13, 1913): 287–94.

Prechel, Harland. *Big Business and the State: Historical Transitions and Corporate Transformation, 1880s–1990s.* Albany: State University of New York Press, 2000.

"Prescription for a Dying Lake." *Science News* 93, no. 7 (February 17, 1968): 159–60.

Pyne, Stephen J. *Fire: A Brief History.* Seattle: University of Washington Press, 2001.

Quinn, Daniel. *The Story of B: An Adventure of Mind and Spirit.* New York: Bantam, 1997.

Reed, George Irving, ed. *Bench and Bar of Ohio: A Compendium of History and Biography.* Vol. 2. Chicago: Century Publishing and Engraving, 1897.

Regier, H. A., and W. L. Hartman. "Lake Erie's Fish Community: 150 Years of Cultural Stresses." *Science* 180 (1973): 1248–55.

Report of the Commissioner of Corporations on the Petroleum Industry, Part II: Prices and Profits. Washington, DC: Government Printing Office, 1907.

Richards, L. A., ed. *Diagnosis and Improvement of Saline and Alkali Soils*. Washington, DC: US Department of Agriculture, 1954.

Richter, Daniel K. *Facing East from Indian Country: A Native History of Early America*. Cambridge, MA: Harvard University Press, 2001.

Rockefeller, John D. *Random Reminiscences of Men and Events*. New York: Doubleday, Page, 1909.

Rodgers, Daniel T. *Atlantic Crossings: Social Politics in a Progressive Age*. Cambridge, MA: Harvard University Press, 1998.

Rose, William Ganson. *Cleveland: The Making of a City*. Cleveland: World Publishing, 1950.

Rosenberg, Charles E. *The Cholera Years: The United States in 1832, 1849, and 1866*. Chicago: University of Chicago Press, 1962.

Rousmaniere, John, ed. *The Enduring Great Lakes*. New York: W. W. Norton, 1979.

Roy, William G. *Socializing Capital: The Rise of the Large Industrial Corporation in America*. Princeton, NJ: Princeton University Press, 1997.

Runte, Alfred. *National Parks: The American Experience*. 4th ed. New York: Taylor Trade Publishing, 2010.

Sabin, Paul. *Crude Politics: The California Oil Market, 1900–1940*. Berkeley: University of California Press, 2005.

Sabin, Paul. "'A Dive into Nature's Great Grab-Bag': Nature, Gender and Capitalism in the Early Pennsylvania Oil Industry." *Pennsylvania History* 66, no. 4 (Autumn 1999): 472–505.

Safire, William. *Safire's Political Dictionary*. New York: Oxford University Press, 2008.

Sampson, Anthony. *The Seven Sisters: The Great Oil Companies and the World They Shaped*. New York: Viking Press, 1975.

Santiago, Myrna I. *The Ecology of Oil: Environment, Labor, and the Mexican Revolution, 1900–1938*. New York: Cambridge University Press, 2006.

Scheiber, Harry N. *Ohio Canal Era: A Case Study of Government and the Economy, 1820–1861*. Athens: Ohio University Press, 1969.

Scott, James C. *Seeing like a State: How Certain Schemes to Improve the Human Condition Have Failed*. New Haven: Yale University Press, 1998.

Segall, Grant. *John D. Rockefeller: Anointed with Oil*. New York: Oxford University Press, 2001.

Shaeffer, C. W. *History of East Providence Refinery*. New York City: E. M. Applegit, 1950.

Shapiro-Shapin, Carolyn G. "'A Really Excellent Scientific Contribution': Scientific Creativity, Scientific Professionalism, and the Chicago Drainage Case, 1900–1906." *Bulletin of the History of Medicine* 71, no. 3 (Fall 1997): 386–411.

Slotkin, Richard. *The Fatal Environment: The Myth of the Frontier in the Age of Industrialization, 1800–1890*. Norman: University of Oklahoma Press, 1998.

Smith, Adam. *An Inquiry into the Nature and Causes of the Wealth of Nations*. Edinburgh: Arch. Constable, 1806.

Smith, Henry Nash. *Virgin Land: The American West as Symbol and Myth*. Cambridge, MA: Harvard University Press, 1950.

Smith, Joseph P., ed. *History of the Republican Party*. Chicago: Lewis Publishing, 1898.

Smith, Stanford H. "Pushed toward Extinction: The Salmon and Trout." In *The Enduring Great Lakes*, edited by John Rousmaniere, 34–45. New York: W. W. Norton, 1979.

Spalding, Heman, and Herman Bundesen. "Control of Typhoid Fever in Chicago." *American Journal of Public Health* 8, no. 5 (May 1918): 358–62.

Speight, James G. *The Chemistry and Technology of Petroleum*. 3rd ed. New York: Marcel Dekker, 1999.

Steinberg, Theodore. *Down to Earth: Nature's Role in American History*. New York: Oxford University Press, 2002.

Steinberg, Theodore. *Nature Incorporated: Industrialization and the Waters of New England*. Amherst: University of Massachusetts Press, 1991.

Steinberg, Theodore. *Slide Mountain: Or, the Folly of Owning Nature*. Berkeley: University of California Press, 1996.

Stoermer, Eugene F. "Bloom and Crash: Algae in the Lakes." In *The Enduring Great Lakes*, edited by John Rousmaniere, 13–20. New York: W. W. Norton, 1979.

Stoll, Steven. *The Great Delusion: A Mad Inventor, Death in the Tropics, and the Utopian Origins of Economic Growth*. New York: Hill and Wang, 2008.

Swartz, U. G. "Some Early Days of Whiting Refinery." *Stanolind Record* 4(9): 11–14, 4(10): 13–17, 4(11): 9–13, 4(12): 13–16.

Tarbell, Ida M. *All in the Day's Work: An Autobiography*. Champaign: University of Illinois Press, 2003.

Tarbell, Ida M. *The History of the Standard Oil Company*. 2 vols. Gloucester: Peter Smith, 1963.

Tarr, Joel A. *The Search for the Ultimate Sink: Urban Pollution in Historical Perspective*. Akron: University of Akron Press, 1996.

Taylor, Alan. "'Wasty Ways': Stories of American Settlement." *Environmental History* 3, no. 3 (July 1998): 291–310.

Taylor, Alice. *Quench the Lamp*. New York: St. Martin's Press, 1990.

Taylor, Graham Romeyn. *Satellite Cities: A Study of Industrial Suburbs*. New York: D. Appleton, 1915.

Tenner, Edward. *Why Things Bite Back: Technology and the Revenge of Unintended Consequences*. New York: Vintage Books, 1997.

Thaler, Richard H. "Toward a Positive Theory of Consumer Choice." *Journal of Economic Behavior and Organization* 1, no. 1 (1980): 39–60.

Thomas, Moyer D. "Effects of Air Pollution on Plants." In World Health Organization, *Air Pollution*, 233–78. New York: Columbia University Press, 1961.

Thomson, Captain J. H., and Boverton Redwood. *Handbook on Petroleum.* London: Charles Griffin, 1901.

Thwaites, Reuben Gold, ed. *Early Western Travels.* Vols. 1–19. Cleveland: Arthur H. Clark, 1904, 1905.

Tocqueville, Alexis de. *Democracy in America.* Translated by Elizabeth Trapnell Rawlings. Boston: Bedford/St. Martin's, 2009.

Townshend, Henry H. *New Haven and the First Oil Well.* New Haven: Privately printed, 1934.

Trachtenberg, Alan. *The Incorporation of America: Culture and Society in the Gilded Age.* New York: Hill and Wang, 1982.

Van Tassel, David D., and John J. Grabowski, eds. *The Encyclopedia of Cleveland History.* Bloomington: Indiana University Press, 1987.

Warner, Hoyt Landon. *Progressivism in Ohio, 1897–1917.* Columbus: Ohio State University Press, 1964.

Warner, Sam B., Jr. *Streetcar Suburbs: The Process of Growth in Boston, 1870–1900.* Cambridge, MA: Harvard University Press, 1962.

Warren, Louis. *The Hunter's Game: Poachers and Conservationists in Twentieth-Century America.* New Haven: Yale University Press, 1999.

Wheeler, Robert A., ed. *Visions of the Western Reserve: Public and Private Documents of Northeastern Ohio, 1750–1860.* Columbus: Ohio State University Press, 2000.

Whipple, George C. *Report on the Quality of the Water Supply of the City of Cleveland, Ohio.* Cleveland, OH: Division of Water Works, 1905.

White, Gerald T. *Formative Years in the Far West: A History of Standard Oil Company of California and Predecessors through 1919.* New York: Appleton-Century-Crofts, 1962.

Whitten, David O., and Bessie E. Whitten. *The Birth of Big Business in the United States, 1860–1914: Commercial, Extractive, and Industrial Enterprise.* Westport, CT: Praeger, 2006.

Wiebe, Robert H. *The Search for Order, 1877–1920.* New York: Hill and Wang, 1967.

Williams, Charles Richard, ed. *Diary and Letters of Rutherford Birchard Hayes.* Vol. 3. Columbus: Ohio State Archaeological and Historical Society, 1924.

Williamson, Harold F., and Arnold R. Daum. *The American Petroleum Industry: The Age of Illumination, 1859–1899.* Evanston, IN: Northwestern University Press, 1959.

Wilson, Ella Grant. *Famous Old Euclid Avenue of Cleveland*. Vol. 2. Cleveland: privately published, 1937.

Wines, Richard A. *Fertilizer in America: From Waste Recycling to Resource Exploitation*. Philadelphia: Temple University Press, 1985.

Wood, Gordon S. *The Creation of the American Republic, 1776–1787*. Chapel Hill: University of North Carolina Press, 1969.

Woolson, Constance F. "Round by Propeller." *Harper's New Monthly Magazine* 45, no. 268 (September 1872): 518–33.

Workers of the Writers' Program of the Work Projects Administration, Indiana (WPAIN). *The Calumet Region Historical Guide*. Reprint. New York: AMS Press, 1939.

Worster, Donald. *Shrinking the Earth: The Rise and Decline of American Abundance*. New York: Oxford University Press, 2016.

Yergin, Daniel. *The Prize: The Epic Quest for Oil, Money, and Power*. New York: Touchstone, 1991.

Zaslowsky, Dyan, and T. H. Watkins. *These American Lands: Parks, Wilderness, and the Public Lands*. Washington, DC: Island Press, 1994.

Zivich, Edward A. *From Zadruga to Oil Refinery: Croation Immigrants and Croatian-Americans in Whiting, Indiana, 1890–1950*. New York: Garland Publishing, 1990.

Žižek, Slavoj. *The Sublime Object of Ideology*. London: Verso, 1989.

INDEX

air pollution, 8, 84, 88; carbon monoxide as, 8, 76, 80, 82, 106; coal sources of, 5, 76, 80, 88, 93, 104, 107, 122, 145n5; drilling sources of, 30; early regulation of, 74–75, 80–82, 84–85, 94–95, 116; effects on plant life, 8–9, 79–81, 93; geography of, 8, 32, 74–75, 86–91, 107; hydrogen sulfide, 106–7; oil sources of, 5, 76–78, 80, 82–83, 91, 93–94, 104, 106–7; sulfur dioxide as, 8, 76, 80

Akron, OH, 15

Albright, Horace, 127

alewife, 113–14

Allegheny River, 33

American Oil Company (AMOCO). *See* Standard Oil of Indiana

Andrews, Samuel, 21–22, 25, 43, 56, 86–87

Anthony, Susan B., 39–40

Appalachian Mountains, 13, 133n6

Archbold, John, 125

Argand, François Pierre Ami, 40

Armour, Philip Danforth, 11, 98

Arthur, J. C., 79–80

Ashmun, G. C., 85

Barstow, F. Q., 42–43

Bayonne, NJ, 94, 97, 143n31, 149n2

Bennehoff Run, 31, 137n8

Berry Lake, 100, 107

Bissell, George H., 19

Black, Brian, 29, 32, 131n1, 135n14

Bolton, Charles E., 91

Boston, MA, 39, 45, 50, 90, 145n2

Bowditch, Ernest W., 90

Briscoe Center for American History, 9

British Petroleum (BP), 3, 10

Brooklyn, NY, 97

Buffalo, NY, 76

Buhrer, Stephen, 62

Bullock, William, 15

Burns, Randall W., 105

Burton, G. B., 42

California, 18, 118, 124, 132n10, 142n12

Canada, 121

capitalism, 4–5, 10, 12, 25, 64, 115, 156n15, 157n15; alienation and, 27, 84, 129; deindustrialization of, 7; depression of 1893, 34, 69–70; externalities of, 37, 44, 54–55, 64, 68–69, 72, 79–80, 111, 128; and labor upheaval of 1877, 49, 62–64, 103; laissez-faire attitudes toward, 8, 23, 29, 45, 47, 49–51, 57–59, 62, 78–79, 83, 92, 114–15, 122; legal framework of, 11–12, 58–59, 93, 106, 119–20, 123; and Panic of 1873, 62; "race to the bottom," 49, 116, 120; resistance to, 64, 69, 79, 117, 121–22

Carnegie, Andrew, 11, 101

Case School of Applied Science, 82

Cassels, J. Lang, 57

Chandler Jr., Alfred, 126, 133n1

Chandler, C. F., 47

175

on, 21; flushing scheme for, 66–67; as navigable channel, 27. *See also* water pollution.

Davison, Eliza. *See* Rockefeller, Eliza Davison
Daykin, A. G., 93
Des Plaines River, 104
Doan Brook, 87–88
Dodd, S. C. T., 120
Drake, Edwin, 19–20, 30, 34–35, 49, 78

East Chicago, In., 101, 110
Edison, Thomas, 122
efficiency, 8, 10, 27, 36, 123, 128; atmospheric limits of, 74, 79–80, 95; fire limits of, 37, 42; frontier attitude toward nature, 14, 88, 95, 116; land limits of, 96–97; scientific 5, 12, 37, 96–98, 124, 126; technological limits of, 68–69, 70–72, 74, 110, 115, 128–29; water limits of, 54, 64, 96, 109, 112, 114–15
Egerton, Frank N., 112
elements, 4, 7, 96. *See also* commodities
Ellsworth, Henry Leavitt, 14
Elmore, Bartow J., 5, 132n4
England, 13, 46–47, 58, 82, 86
environmental justice, 4, 131n1. *See also* efficiency
environmental limits, 6, 37, 42, 115, 128. *See also* efficiency
Environmental Protection Agency (EPA), 116
environmentalism; "free market," 10, 129
Euclid Avenue. *See* Millionaires' Row
Eveleth, J. G., 19
Exxon/Mobil, 3, 128; merger of, 6, 124. *See also* Standard Oil of New Jersey; Standard Oil of New York

Farley, John H., 81
Federal Water Pollution Control Act (1948), 112–14
Fernow, B. E., 79–80
fire: burn and flash points, 31, 41–44, 47; from exploding lamps, 45–48, 52, 121; insurance costs of, 8, 36–37, 45; at oil wells, 30–31; price wars and, 7, 41–42; at refineries, 37–39, 46, 48; spontaneous combustion and, 5, 7, 31, 46, 48; telegraphic alarm boxes, 36–37; transportation hazards of, 32–34; workplace safety and, 7, 28, 33–34
Flagler, Henry, 25, 34, 56, 86, 120
Ford, Henry, 4
Forest Hill estate, 26, 69, 86
Foster, Charles W., 24–25
Foster, W. H., 74, 81
fracking (hydraulic fracturing), 10
Frank, Alison, 9
Franklin, Benjamin, 13–15, 85
Frasch, Herman, 97–98, 106

Gary, Albert H., 102
Gary, IN, 102, 104–5, 108, 110–11, 116–17
gasoline, 3, 35, 47, 104; as danger, 7, 39, 52; early use as fuel, 28–29, 44–45, 51, 132n7; as waste, 8, 10, 44
Gavagan, Thomas, 94
Gehrke, Betty, 114
Gesner, Abraham, 56
Gilded Age, 4–5, 10, 12, 23, 29, 35, 38, 40, 49–50, 59, 61, 66, 69, 71–72, 74, 78, 80, 83, 89, 92, 94–95, 128; "long," 6, 9, 96
Golden, Peter G., 28–29, 35, 44, 51, 123
Gould, Jay, 11
Grand Calumet River, 101–2, 104, 110–11

public perception of, 65, 77, 84, 94, 104, 109, 114–15; public relations efforts of, 125–27; sales agents of, 41–44, 120–21; statistical department, 23; trust agreement and, 24, 52, 96, 119–20
Standard Oil of California, 124
Standard Oil of Indiana, 9–10, 124
Standard Oil of New Jersey, 120, 124
Standard Oil of New York, 23, 120, 124, 126–27
Stanton, Elizabeth Cady, 39–40
Stein, Murray, 114
Stoll, Steven, 129
Swartz, U. G., 107–8

tank car (railroad), 5, 33–34
Tarbell, Ida, 125
Taylor, Alice, 40
Texaco, 123
Texas, 9, 123
Thompson, William P., 32, 37, 42
Thomson, Captain J. H., 46–47
Time, 114
Titan, 15
Titusville, PA, 30–31, 41
Trachtenberg, Alan, 12
turpentine, 18, 27, 29–30, 119

Union National Bank of Cleveland, 89–90
United States Congress: early representatives from Ohio, 15; and kerosene fire test, 47–48
United States Department of Agriculture (USDA), 79, 95
United States Supreme Court, 3, 6, 106, 123–25
University of Chicago, 105, 107, 116
University of Indiana, 109

University of Texas, 9
US Public Health Service, 110–11
U.S. Steel, 101–2, 108, 111, 121

Vandevelde, John, 81–82
Vaseline, 29, 104, 119
vertical integration, 4–5, 21–22, 29, 67, 103, 124
Vial, John, 61
Visions of the Western Reserve, 14

Walworth Run, 35, 61, 78–79, 82–84
Washington, DC, 116
waste treatment, 55, 72, 109–11; ozone as, 110–11, 115
water pollution, 114, 116; acid waste sources of, 5, 8, 10, 55–56, 58–61, 67–68, 111; American jurisprudence and, 58–59; in Chicago, 104, 109, 111; in Cleveland, 23, 53–55, 57–58, 60–61, 66–67, 70–73; and diseconomies, 68–69; in drinking water, 10, 54–55, 104, 109–11, 128; gasoline sources of, 10, 44; organic sources of, 55, 57–58, 60, 66–67, 71–72, 83–84, 104, 110; petroleum sources of, 57–58, 60, 62, 64, 71–72, 78–79, 82–83, 107, 110–13, 128; steel mill sources of, 103, 110
waterworks: water intake crib and, 8, 54, 68–72, 104, 109–11; in Whiting, 99–100
Waukegan, IL, 111
Wealth Against Commonwealth, 121
Western Reserve (of Connecticut), 14
Western Reserve University, 89
Western Union, 36
whale oil, 17–18, 20, 27, 29, 41, 45–46, 97
Wheeler, Robert A., 14